Key Thinkers in Neuroscience

Key Thinkers in Neuroscience provides insight into the life and work of some of the most significant minds that have shaped the discipline. Studies of the human brain and nervous system have been varied and complex, and there have been many pioneers, including a number of nobel laureates, who have broken new ground in neuroscience research.

Adopting a chronological and multi-disciplinary approach to each Key Thinker, the book highlights their extraordinary contributions to neuroscience. Beginning with Santiago Ramón y Cajal and concluding with Patricia Churchland and Paul Churchland, this book provides a comprehensive look at the new ideas and discoveries that have shaped neuroscientific research and practice, and at the people who have been invaluable to this field.

This book will be an indispensable companion for all the students of neuroscience and the history of psychology, as well as for anyone interested in how we have built our knowledge of the brain.

Andrew P. Wickens is a former Senior Lecturer in Psychology and Neuroscience at the University of Central Lancashire, UK. His main area of expertise is in biological psychology and neuroscience.

Key Thinkers in Neuroscience

Andrew P. Wickens

Routledge
Taylor & Francis Group
LONDON AND NEW YORK

First published 2019
by Routledge
2 Park Square, Milton Park, Abingdon, Oxon OX14 4RN

and by Routledge
711 Third Avenue, New York, NY 10017

Routledge is an imprint of the Taylor & Francis Group, an informa business

British Library Cataloguing-in-Publication Data
A catalogue record for this book is available from the British Library

Library of Congress Cataloging-in-Publication Data
Names: Wickens, Andrew P., author.
Title: Key thinkers in neuroscience / Andy Wickens.
Description: Abingdon, Oxon ; New York, NY : Routledge, 2019.
Identifiers: LCCN 2018020814| ISBN 9781138576124 (hardback : alk. paper) | ISBN 9781138576131 (pbk. : alk. paper) | ISBN 9781351271042 (ebook)
Subjects: LCSH: Neuroscientists—Biography. | Neurosciences—History.
Classification: LCC QP353.3 .W53 2019 | DDC 612.8/233—dc23
LC record available at https://lccn.loc.gov/2018020814

ISBN: 978-1-138-57612-4 (hbk)
ISBN: 978-1-138-57613-1 (pbk)
ISBN: 978-1-351-27104-2 (ebk)

Typeset in Sabon and Gill Sans
by Florence Production Ltd, Stoodleigh, Devon, UK

Contents

Preface ix
Acknowledgements xiii
Dedication xiv

1 Santiago Ramón y Cajal (1852–1934) 1

2 Charles Scott Sherrington (1857–1952) 8

3 John Newport Langley (1852–1925) 16

4 Korbinian Brodmann (1868–1918) 23

5 Wilder Graves Penfield (1891–1976) 30

6 Edgar Douglas Adrian (1889–1977) 37

7 Henry Hallet Dale (1875–1968) 44

8 Walter Rudolf Hess (1881–1973) 51

9 Karl Spencer Lashley (1890–1958) 58

10 Giuseppi Moruzzi (1910–1986) and Horace Winchell Magoun (1907–1991) 65

11 Donald Olding Hebb (1904–1985) 72

12 Nathaniel Kleitman (1895–1999) 79

13 Andrew Lloyd Hodgkin (1914–1998) and Andrew Fielding Huxley (1917–2012) 86

14 Paul Donald MacLean (1913–2007) 93

15 John Carew Eccles (1903–1997) 100

16 Bernard Katz (1911–2003) 107

17 Seymour Solomon Kety (1915–2000) 114

18 Brenda Milner (b. 1918) 121

19 James Olds (1922–1976) 128

20 Julius Axelrod (1912–2004) 135

21 Rita Levi-Montalcini (1909–2012) 142

22 Roger Wolcott Sperry (1913–1994) 149

23 David Hunter Hubel (1926–2013) and Torsten Nils Wiesel (b. 1924) 156

24 Arvid Carlsson (1923–2018) 163

25 Norman Geschwind (1926–1984) 170

26 Paul Greengard (b. 1925) 177

27 Benjamin Libet (1916–2007) 184

28 Eric Kandel (b. 1929) 191

29 Bert Sakmann (b. 1942) and Erwin Neher (b. 1944) 198

30 David Coutnay Marr (1945–1980) 205

31 Mortimer Mishkin (b. 1926) and Leslie Ungerleider (b. 1946) 212

32 Timothy Bliss (b. 1940) and Terje Lømo (b. 1935) 219

33 Richard Axel (b. 1946) and Linda Buck (b. 1947) 226

34 Giacomo Rizzolatti (b. 1937) 233

35 John O'Keefe (b. 1939) 240

36 Michael Gazzaniga (b. 1939) 247

37 Joseph LeDoux (b. 1949) 254

38 Patricia Churchland (b. 1943) and Paul Churchland (b. 1942) 261

Author index 269
Subject index 273

Preface

At the end of 2016, I left my university position after 27 years. I had grown weary of sitting in front of a computer, marking essays and filling in feedback forms – a pursuit that is far more mentally draining than writing a book. It is sometimes said that when one door shuts, another opens. But in my case, I had not quite bargained how much I would miss having a forced work routine that gave at least some sort of structure and purpose in my life – even if I disliked the stress of working for an institution seemingly obsessed with increasing the work load of its staff year by year. It was therefore a relief after some months of frittering away my time that I was offered the chance by Ceri McLardy at Taylor and Francis to write *Key Thinkers in Neuroscience*. It was a truly exciting opportunity.

My first task was to compile a list of neuroscientists who had made significant key discoveries in their respective fields. I quickly drew up a list of some forty names over a cup of coffee. Admittedly, not all the original names have stood the test of my scrutiny, but the majority have remained in place. I realise no two people would fully agree on a list, and there cannot be a definitive roll call of the greatest key thinkers. Everybody will have different opinions, and at the end of the day my choices are bound to be partly subjective, reflecting the interests and biases of the compiler. Despite this, I think the majority of the neuroscientists I have covered in this book are not contentious. It was also immediately obvious to me that *Key Thinkers* would work best if the names were presented in a chronological framework, and in doing so provides a brief if somewhat fragmented modern history of neuroscience.

In some respects, writing a book is not unlike composing a musical score, and in this regard I was particularly keen to start my book on a strong major chord, with a figure whose presence can be felt throughout all the chapters. And to this end, there was only one person who fitted the bill – Santiago Ramón y Cajal, who among his many remarkable achievements was the main instigator of the neuron theory on which modern day neuroscience is firmly anchored. Some may quibble by arguing I should have begun with one of his predecessors – perhaps Camillo Golgi, since he invented the silver staining technique that Cajal hijacked for his own studies. Others may point to David Ferrier for his work on cortical localisation or even Jean-Marie Charcot for his histological-anatomical method that helped finding the field of neurology. All three have their merits – and their weaknesses. Golgi was a staunch supporter of the reticulum theory of nervous organisation, Ferrier believed the angular gyrus was the site of the primary visual cortex and Charcot's great work was humbled by his conviction that hysteria was an organic disorder. No, it has to be Cajal who is the father of modern neuroscience and in my eyes the greatest of them all.

It was only after I had put together my original list of names that I began to consider more carefully what is meant exactly by *key thinker*. The answer to this question is perhaps not as simple as it first appears. After some thought, I came to the conclusion that a key thinker is someone whose ideas have made a vital *conceptual* contribution to understanding the brain and nervous function while also having a profound impact on the *future development* of neuroscience. This more precisely defined concept of key thinker led me to jettison some of the original names from my list. These included Hans Berger, Godfrey Hounsfield and Seiji Ogawa. All of these three men have pioneered great technological advances (electroencephalograph [EEG], computed axial tomography [CAT] and functional magnetic resonance imaging [fMRI], respectively) which have opened up new areas of neuroscience, and there is no doubt that their impact on the brain sciences has been colossal. Yet, none of these three men can be considered as a true key thinker in neuroscience, despite them opening up new avenues, perhaps even entire fields, for investigation.

Despite this, key thinkers come in many guises. In drawing up a list of key thinkers, I was initially aware (or so I believed) of a division between those who have made fundamental contributions to an understanding of how the nervous system operates (e.g. Sherrington, Dale and Adrian) and those who have attempted to explain how the nervous system gives rise to behaviour such as emotion, memory and thought (e.g. LeDoux, Kandel and Gazzaniga). It was also beguiling to think that the former are represented by the neuroscientists of yesteryear and the latter by more recent contemporary neuroscientists. But on closer scrutiny, it is obvious that this is not the case. We may believe that we know all there is about how the nervous system operates, from the action potential, to neurotransmission, to second messengers and so on. But, really, this is fabrication. There is still much more to know (as Kandel's work on *Aplysia* shows) – and who knows, one day even our fundamental understanding of nervous action may be radically overhauled. But when we come to understanding the functions of the brain and its impact on behaviour, then the facts are simple: at present, we know very little.

There are a couple of people in my list of key thinkers that may raise an eyebrow. One is Benjamin Libet who undertook pioneering work of human consciousness, free will and the initiation of action. Although much of his work is controversial, many of the questions he raised are important and remain unanswered. He also happens to be one of my favourite neuroscientists. I also wanted to end my book by looking into the future, and to this end I firmly believe that one day neuroscience will help solve some of the big philosophical problems about the mind. Consequently, I have chosen the philosophers Patricia and Paul Churchland, who have done much to establish neurophilosophy as a branch of neuroscience. I have to confess I am playing devil's advocate here since I don't endorse their belief in eliminative materialism (I am much more sympathetic to non-reductive materialism). But this does not matter. The nature of the mind remains the greatest unsolved mystery of all. As the philosopher Daniel Dement once remarked, it is so mysterious, we don't even know how to think about it yet. But when we do, I am sure it will be a neuroscientist who unties the Gordian knot.

One issue that niggled me as I wrote this book was the lack of women. In fact, there are just six (out of a total of forty-six). The reasons for the lack of women in the neurosciences should be more widely debated (this is not the place) and certainly more done to rectify the situation. Whatever the reason, I am convinced women find the subject of brain science just as fascinating as men, and I hope the situation is changing. Although the number of women in my book is small, each should be a true inspiration to any aspiring female student. As I write these words, two have won Nobel Prizes and another is approaching her hundredth birthday and still going strong.

I hope this book is easy to read while still maintaining its academic rigorousness. But most of all I want *Key Thinkers in Neuroscience* to inspire. For whatever reason, despite my best attempts, I never made it as a famous neuroscientist, and being a writer is very much a poorer option (and sadly not respected in today's universities). But the people I write about here are my heroes – and I hope their greatness will be recognised as such by the readers of this book. Neuroscience is a very young discipline, which makes it a very exciting one too. It has been said that the amount of science and technological advancement is increasing exponentially. If we assume that a doubling of our knowledge is taking place every 10 years (a conservative estimate), the next 50 years should see a stunning revolution in how we think about the brain. The student of neuroscience is fortunate in becoming acquainted with the subject at this very exciting time. If I can get some of this excitement across, then this book will have fulfilled its task. But, even better . . . if you, as a reader of this book, was featured in its next volume, then this would have pleased me the most.

Andrew Wickens
Easter Monday, 2018

Acknowledgements

First and foremost I would like to thank Ceri McLardy for offering me this most wonderful of opportunities to write *Key Thinkers in Neuroscience*. It not only came at a propitious time for me but has also proven a great joy to write. Although I have tried to keep my correspondence to a minimum, there have been times when I have requested information from other people, and in this respect I would especially like to thank professors Jim Olds, Michael Gazzaniga and Joseph LeDoux, who replied promptly to my request for titbits of information. Even a simple email reply can sometimes be a great thrill (especially from a legend). I would also like to thank Catherine Lemelin at Amherst College, and David Kahn at Central High School, Philadelphia, for helping me with my enquiries. Although I am no longer a lecturer at Uclan, I still keep in close contact with the Psychology Department, and I would like to acknowledge Linden Ball for his assistance – not least for allowing me to keep my Uclan email account (an important acquisition in today's digital age). Still with Uclan, many thanks to Nikola Bridges and Cassie Richardson for spending some of their time in a very boring exam board meeting, helping me come up with some female names for this book. Finally, I would like to thank Kathy Wright for her love and support. Being a typical bloke, I rarely tell her how much I appreciate it.

Dedication

In 1973, when I was 14 years old, as a raw and uncouth teenager, I heard *Dark Side of the Moon* for the first time. It was so utterly mesmerising and overwhelming that I felt like I had been transported to another planet. Maybe I had and have never quite returned. Without it, I think I would have gone on to study literature or geography. Instead, I developed an interest in psychology and then neuroscience. This may be a silly and pointless thing to do (although it is not to me) . . . but I dedicate my thanks to Pink Floyd.

Chapter 1

Santiago Ramón y Cajal (1852–1934)

If there is one person above all others who is the father of modern neuroscience, then it is surely Santiago Ramón y Cajal, whose work led to the formulation of the neuron doctrine, which holds that nerve cells form the fundamental structural and functional units of the nervous system.

The early portends for the father of modern neuroscience were not promising. The son of a struggling country doctor, born in the small town of Petilla de Aragon in north eastern Spain, Cajal was a headstrong and rebellious youth, expelled from school and often punished for bad behaviour. This included an act of vandalism at the age of 11 when he blew a hole in his neighbour's gate with home-made explosive! Although recalcitrant, Cajal had a passion for drawing and dreamed of becoming a painter, although he was first apprenticed as a shoe-maker and then a barber. Fortunately, his father managed to turn his wayward son's interest to anatomy after teaching him about the human skeleton using bones that had allegedly been unearthed from a local cemetery. Cajal would later confess, "Henceforth, I saw in the cadaver, not death . . . but the marvellous workmanship of life". This formative experience encouraged him to begin medical training at the nearby University of Zaragoza. After graduating in 1873, Cajal served in Cuba, and then under the Spanish rule, he served as a military doctor. However, he would spend less than 2 years in Cuba before contracting malaria and tuberculosis. In a state of frail health, Cajal went back to Spain intent on changing vocation.

In 1875, Cajal returned to the University of Zaragoza as assistant professor – a position enabling him to study for his doctorate. Two years later after passing his examinations in Madrid, Cajal spotted a modest microscope for sale in a shop, with a magnification sufficient for his research interests. Using every peseta he had saved from Cuba, Cajal purchased the instrument and took it home where he set up a laboratory to examine the pathogenesis of inflammation. At the time, Cajal was still in poor health and suffered a pulmonary haemorrhage in 1878. Nonetheless, his investigations, some of it examining how nerve fibres terminated in muscle, led him to be appointed as director of the anatomical museum at the University of Zaragoza in 1879. He then rose to chair of anatomy at the University of Valencia in 1883 – a period that also saw Cajal publishing his first scientific papers. Interestingly, while his full surname was Ramón y Cajal, he dropped the surname of his father (Ramón) and tended to publish his work simply as Cajal (the surname of his mother).

At the age of 35, in 1887, Cajal moved to the University of Barcelona. It proved to be a turning point in Cajal's research life, for in the same year he was shown samples of nerve cells stained with the silver impregnation method. This technique invented by the Italian Camillo Golgi in 1873 was one that only stained a small percentage of neurons (around 2%) in any preparation of nervous material. But far from being a disadvantage, this allowed the few affected neurons to stand out in black against a silvery-yellow background. Now, the cell's morphological features such as their dendrites could be viewed in unsurpassed clarity and axons followed through the nervous morass. Enthralled by what he saw, Cajal vowed to learn the technique, although his first attempts to do so were disappointing since the results were capricious and unreliable. But with a little experimentation, Cajal realised that if he stained the tissue twice, this markedly improved the reliability and effectiveness of the technique. And even better results were obtained if he stained unmyelinated axons – a finding that prompted him to use his technique on young animals.

At this point, Cajal engaged in a feverish burst of research – staining many areas of the brain and drawing his observations. Realising this work was providing many new findings, he began publishing his own journal, with copies sent to the most important neuroanatomists of Europe. The inaugural copy appeared in 1888, with six articles by Cajal, in a format that was to set new standards for reporting neuroanatomical findings. The first paper, for example, described the anatomy of the bird's cerebellum. It showed the cerebellum to be a laminated structure with the axons of the small stellate cells in one layer ending freely over the cell bodies of the much larger Purkinje cells. Cajal also described the mossy fibres for the first time, which entered the cerebellum from an outside site and formed a relationship with the small dendrites of the granule cells. But Cajal's real genius lay in the way he depicted these neuronal arrangements. From his drawings, the reader could see the neural structure of the cerebellum at a glance. In fact, his illustrations were a composite of many different drawings, based on numerous hours of meticulous microscopic examination and expressed with some degree of artistic imagination from Cajal's own anatomical knowledge. Nonetheless, they captured the structural complexity of the cerebellum in exquisite detail. Even today, it can be argued that Cajal's drawings of the nervous system have never been surpassed.

Cajal's work, however, was to have a far bigger theoretical significance. For much of the nineteenth century, the prevailing view of the nervous system was that it formed a reticulum – namely, a continuous network of web-like interconnected or fused fibres. This theory had been developed by Joseph von Gerlach in 1872, although it was supported by Golgi, whose stain Cajal had enthusiastically adopted. But if this theory was correct, then it made the nervous system unique, for it was known all the other tissues of the body were made up of individual cells – a theory credited to Matthias Schleiden and Theodor Schwann in the 1830s. Nonetheless, right from the beginning of his research with the improved Golgi technique, Cajal was unable to find any evidence of axonal fusing, and it soon led him to assert with his characteristic forthrightness that nerve cells should be regarded like any other in the body – namely, independent units. Indeed, his observations showed that axons, instead of fusing, appeared to terminate as "free aborizations" on the dendrites of their recipient cells. It was an observation at odds with Golgi who had proposed that the dendrites only had a nutritive role in keeping the cell alive. A year later, Cajal came up with an even more profound idea when he deduced the direction of information flow through nerve cells. In short, he recognised that the dendrites provided the receptive part of the nerve cell, while the axon was its "emitting" and "distribution" apparatus. Cajal called his theory "the law of *dynamic polarisation*".

Thus, Cajal had laid the foundations for a totally new understanding of the nervous system within a period of less than 2 years.

Yet, despite his valiant attempts, Cajal's work was not attracting attention from others – largely because Spanish was not widely read or spoken by the wider scientific community. In an attempt to address this, Cajal travelled to Berlin in 1889 to present his findings at an international congress. Unable to speak German and only managing a little broken French, Cajal set up his microscopes and invited delegates to view his silver-impregnated slides. One person to accept his invitation was Albert von Kölliker – then the most respected neuroanatomist of his time, who was also a champion of cellular theory and author of several hundred papers on tissue biology. Kölliker was immediately impressed with what he saw, and with his support Cajal was soon able to influence the rest of the neuroanatomy community with his pioneering research and ideas.

Around this time, other investigators were also beginning to cast a shadow on reticulum theory. One such researcher was the Swiss anatomist Wilhelm His. During the early 1880s, His had carefully followed the stages of early nervous system development in tissue taken from aborted human embryos aged between 2 and 8 weeks, and it led to the discovery that the spinal cord and its nerves arose from a primitive type of cell called a "neuropile". But as he watched the axon fibre grow from the neuropile, followed by the emergence of the cell body and dendrites, at no point could he see the axons fuse as Golgi insisted. The same conclusion was also reached by another Swiss researcher, August Forel, who lesioned the motor cranial nerves in rodents and then traced the degeneration of the axons back to the brain. The results showed that the degenerating axons only produced small amounts of damage in small and specific areas of the brain stem. In other words, it was apparent that the axons were not continuous or joined with the neurons they reached in the brain – for if this was the case, then the degeneration should have been far more extensive and extend to many other areas. By 1891, the evidence had become so persuasive that Wilhelm Waldeyer was able to collect sufficient information to compile an extensive review supporting the concept of nerve cells, arguing they formed the main structural, embryological and functional units of the nervous system. In making his case, Waldeyer also referred to nerve cells as *neurons*. It was an inspired choice of term for the new theory would become known as the *neuron doctrine* – a phrase that helped encapsulate a biological principle supporting Cajal's position.

Despite this, the neuron doctrine continued to be strongly criticised by the reticulists led by Golgi, who stuck to their belief that communication between axon fibres took place through continuity and not through direct contact. Indeed, the contact hypothesis was problematical for the neuron doctrine since it was not clear how the nerve impulse (which by now was known to be electrical in nature) could travel from one separate nerve cell to another. In this respect, the Golgi technique offered few clues. The strength of the Golgi stain had been its random selectivity. By only staining a small percentage of nerve cells in any tissue sample, it had made them highly visible for a microscopist to observe. But this was also a weakness when it came to establishing their connections since it was unlikely the stain would highlight two connecting nerve cells. And even if this did occur, the tiny cells of the nervous system meant that visualisation of the contact was beyond the resolution of the microscope of the times. The one exception to this was the much larger neuromuscular junction. Here, the motor fibre and muscle end plate could be seen in greater detail, and in 1886, the German Wilhelm Kühne reported that he could see a clear gap between the two. Kühne conveyed this to the Royal Society of London in 1888, speculating the nerve ending was open in some way, allowing its fluid to escape and cross the tiny gap, before reaching the muscle.

In 1894, Cajal received an invitation from the Royal Society to deliver their annual Croonian lecture. The decision to invite Cajal had come from Charles Sherrington whose main interest concerned the spinal circuitry of reflex action – work that required him to map the pathways of nerves from sensory stimulus to motor response. His attempts to trace the flow of information through the spinal cord had also convinced him that the nervous system was composed of nerve cells whose fibres made contact without fusing. Cajal travelled to London, gave his talk in French and stayed with the Sherrington family for his visit. It is not known if the two men discussed naming the contact point. However, soon after, Sherrington was invited to write a chapter in Sir Michael Foster's *Textbook of Physiology*. He accepted, and in doing so he coined the term *synapsis* (from the Greek meaning "to clasp") to describe the "surface of separation" between axon and nerve cell. The word soon morphed into *synapse*. Sherrington could only speculate how neural information might cross this junction. Nonetheless, he was certain that synapses existed – not least because he knew that the sum speed of reflex activity was much slower than the speed of nervous conduction.

In 1906, Golgi and Cajal were awarded the Nobel Prize in recognition of their work on elucidating the structure of the nervous system. It was also to be the only time the two men met. Sadly, they did not get on. Nor was the situation helped when Golgi in his acceptance speech strongly denounced Cajal's nerve cell theory by telling the audience "it was going out of favour". History has since shown that Cajal was correct in regard to the neuron doctrine and many other aspects of nerve function. In Golgi's defence, it can be said that Cajal's position was not proven in 1906, and for many the concept of a synapse was an implausible concept, leaving important questions about the nature of neural transmission unanswered. In fact, the existence of synapses in the central nervous system (CNS) would not be fully accepted until the invention of the electron microscope in the 1950s. Even then, there were those who opposed the idea of chemical neurotransmission.

If these were Cajal's only achievements, they alone would have placed him in the pantheon of the greatest neuroscientists of all time. But there was much more which this short biography can only touch upon. In the years after he exploited the Golgi stain to describe the structure and function of the nervous system, thereby laying down the modern foundations of neuroscience, Cajal began working on the degeneration and regeneration of nerve fibres in the spinal cord and visual system. During this time, he also wrote several books of great importance, including his magnificent two-volume *Textura del Sistema Nerviosa del Hombre y de los Vertebrados* (*Histology of the Nervous System of Man and Vertebrates*) in 1904, which contained nearly 2000 pages and 887 illustrations – graphically revealing nearly every region of the brain. Over his long research career, Cajal would write almost 300 articles, found two scientific journals, help pioneer colour photography and serve as a director of several institutions. He also had time to write non-scientific books including *Advice for a Young Investigator* (1916) and a colourful biography *Recollections of my Life* (1917). Towards the end of his life, Cajal's fame was so great that the Spanish authorities tried to put his portrait on bank notes and postage stamps (an accolade he refused to permit), although he accepted a life senator's role in Madrid. Retiring from university work in 1922, Cajal continued writing until his death at the age of 83 in 1934. An obituary by the Italian neurologist Ernesto Lugaro eulogised that Cajal had contributed more to the knowledge of neuroscience than all the efforts of his fellow colleagues put together. Few would disagree.

Cajal's major writings

Cajal, S.R. (1899/1904) *Histology of the Nervous System of Man and Vertebrates*. Translated by N. Swanson and L.W. Swanson. New York: Oxford University Press (1995).

Cajal, S.R. (1906) The structure and connexions of neurons. In: *Nobel Lectures, Physiology and Medicine 1901–1921*. New York: Elsevier (1967).

Cajal, S.R. (1913/1914) *Degeneration and Regeneration of the Nervous System*. Translated by R.M. May. London: Oxford University Press (1928).

Cajal, S.R. (1916) *Advice for a Young Investigator*. Translated by N. Swanson and L.W. Swanson. Cambridge, MA: MIT Press (1999).

Cajal, S.R. (1917) *Recollections of my Life*. Translated by E.H. Craigie, with the assistance of J. Cano. Cambridge, MA: MIT Press (1989).

Cajal, S.R. (1933) *Neuron Theory or Reticular Theory? Objective Evidence of the Anatomical Unity of Nerve Cells*. Translated by M. Ubeda-Purkiss and C.A. Fox. Madrid: Instituto Ramón y Cajal (1954).

Further reading

De Carlos, J.A. and Borrell, J. (2007) A historical reflection of the contributions of Cajal and Golgi to the foundations of neuroscience. *Brain Research Reviews*, 55, 8–16.

DeFelipe, J. (2002) Sesquicentennial of the birthday of Santiago Ramón Cajal (1852–2002), the father of modern neuroscience. *Trends in Neuroscience*, 25, 481–484.

DeFelipe, J. (2013) Cajal and the discovery of a new artistic world: The neuronal forest. *Progress in Brain Research*, 203, 201–220.

Jones, E.G. (1994) The neuron doctrine. *Journal of the History of Neurosciences*, 3, 3–20.

López-Muñoz, F., Boya, J. and Alamo, C. (2006) Neuron theory, the cornerstone of neuroscience, on the centenary of the Nobel Prize award to Santiago Ramón y Cajal. *Brain Research Bulletin*, 70, 391–405.

Shepherd, G.M. (1991) *Foundations of the Neuron Doctrine*. New York: Oxford University Press.

Sotelo, C. (2003) Viewing the brain through the master hand of Ramón y Cajal. *Nature Reviews Neuroscience*, 4, 71–77.

Swanson, L.W., Newman, E., Araque, A. and Dubinsky, J. (2017) *The Beautiful Brain: The Drawings of Santiago Ramón y Cajal*. New York: Abrams Books.

Chapter 2

Charles Scott Sherrington (1857–1952)

Sherrington's greatest legacy to neuroscience was his elucidation of the spinal reflex – work that led him to recognise the importance of inhibition in neuronal function, discover proprioception and identify the synapse (a term he coined) as the site where nerve cells interact.

If Cajal was the prime mover in establishing neurons as the basic building block of the nervous system, then it was Sherrington who made the first step in explaining how neurons functioned together as units through his analysis of the reflex. According to official biography, Sherrington was born in London in 1857 and was the son of a country doctor who died during his infancy. There is some doubt over this claim, however, and other accounts say that Sherrington was the illegitimate son of Caleb Rose, a classical scholar and amateur archaeologist, who only got around to marrying his mother Anne Brookes in 1880. Whatever the truth of the matter, Sherrington grew up in Great Yarmouth and Ipswich where Rose worked as a physician. Rose also installed in his step-son a lifelong appreciation for art and literature. From 1871 to 1876, Sherrington went to Ipswich Grammar School and received an education that was strongly classical and mathematical, although he chose medicine when beginning his undergraduate education at St Thomas's Hospital in London in 1876. Three years later, Sherrington combined his clinical work with study at Cambridge where he joined Gonville and Caius College. An able athlete who represented his college at rowing and rugby, Sherrington obtained a first class with distinction in the Natural Science Tripos – an exceptionally broad type of bachelor's degree unique to Cambridge. It was also around this time that Sherrington became a member of a group centred around Sir Michael Foster who had been pivotal in turning Cambridge into a world renounced centre for physiological research. Sherrington obtained his medical degree in 1885, having already managed to publish his first research paper (examining the effects of decortications in dogs on the medulla and spinal cord) with John Newport Langley a year earlier.

After leaving Cambridge, Sherrington's career took a varied course. In the winter of 1884, he joined Friedrich Goltz in Strasbourg where he continued his investigations into nerve tract degeneration following cortical damage. At this point, however, it seems that Sherrington was more interested in infectious diseases, and in the summer of 1885, he took advantage of an opportunity to investigate a cholera epidemic in Spain, where according to some sources he met Ramón y Cajal for the first time.[1] After visiting Italy where another cholera outbreak had occurred, Sherrington worked for a short period in Berlin under the great cellular pathologist Rudolf Virchow and then for a few months with the bacteriologist Robert Koch. This was to prove a valuable experience for Sherrington, who was later able to develop a diphtheria vaccine in 1894, which he used to save the life of his young nephew. In 1887, Sherrington returned home and began a lectureship in physiology at St Thomas's Hospital where he started to take a greater interest in brain pathology.

In 1891, Sherrington moved to the Brown Institution, a veterinary hospital run by the University of London with excellent research facilities, and it was here he initiated his interest in spinal reflexes. More specifically, he began examining the patella reflex, otherwise known as the knee jerk, which is produced by tapping the tendon below the kneecap – a test routinely used by doctors to assess the health of the nervous system. This reflex had first been described by the Germans Carl Westphal and Wilhelm Erb in 1875, but the two men had arrived at different ideas about how it occurred. Westphal believed the reflex was caused by a simple twitch of the quadriceps (the muscle running down the front of the thigh), which was transmitted from the force of the tapped tendon. In contrast, Erb believed the reflex involved the flow of information along a sensory nerve, that entered the spinal cord from the tendon, which was then turned back to contract the quadriceps muscle through a motor pathway.

Sherrington first identified the spinal roots that mediated the patellar reflex and then severed their nerve pathways that relayed information to and from the leg muscles. This proved that Erb was correct: the reflex involved a nervous impulse from the tendon to the spinal cord, which was re-directed back to the quadriceps. But Sherrington found this reflex to be much more complex than it first appeared, because the tendon tap also lessened the muscle tension in the hamstrings that ran along the back of the thigh. In fact, this was the first demonstration, made in 1891, of something Sherrington repeatedly found in all spinal reflexes – namely, the reciprocal action of agonist and antagonistic muscles (i.e. as one muscle contracts the other relaxes). This would lead Sherrington to eventually realise that the concept of a simple excitatory reflex, as had been theorised by prior researchers, did not exist. Instead, a reciprocal response is essential for reflexive action, including the knee jerk, since the reduced tension of the hamstrings is necessary to allow the quadriceps to stretch. From this, Sherrington deduced that the knee jerk must involve two motor neurons: one producing excitation of the quadriceps and the other causing inhibition of the hamstrings.

In 1895, Sherrington was appointed as chair of physiology at University College, Liverpool. Here, he started in earnest a decade of research mapping (mainly in cats and monkeys) the sensory and motor flow of the spinal nerves[2] – work that he had begun in London and a necessary prerequisite for understanding the neural circuitry of more complex reflexes. To identify the destination of the motor nerves, Sherrington typically anaesthetised an animal and used a mild electric current to stimulate the ventral roots. By doing this, Sherrington discovered that most muscles of the body are actually innervated from several roots at different levels of the spinal cord. To classify the sensory pathways passing back into the spinal cord, however, Sherrington prompted reflexive action by pinching areas of skin and then assessed the extent to which this action was abolished after cutting the dorsal roots. By using this approach, Sherrington was able to map the body's territories of touch that fed into each spinal nerve. It showed that several overlapping skin fields often fed into one single dorsal root.

Sherrington admitted he found such research laborious, dull and pedestrian, but it led to what is arguably his greatest discovery. Prior to Sherrington, it was believed that the skeletal muscles only produced movement through their ability to contract. Moreover, there was no reason to suppose muscles acted as sensory organs since they did not appear to contain any types of receptor. Thus, it was assumed the spinal cord and brain received all of their sensory input from the skin and joints. However, when Sherrington cut the ventral spinal roots (a procedure, in theory, that should have destroyed all motor nerves innervating the muscles), he found nearly half of the fibres were "untouched" in the outflow pathways. The only plausible explanation was that these were fibres originating in the muscles instead and entering the spinal cord. But if this was the case, where were the sense organs? Sherrington's investigations soon led to him identify elongated bundles of fibrous threads embedded in the muscle tissue, called "muscle spindles", which detected the amount of stretch taking place in the muscle. This mechanism was shown to be important for instigating a compensatory reflex contraction when the muscle became stretched by a heavy load. In 1906, Sherrington called this response "proprioception" (from the Latin proprius meaning "one's own"). He also regarded proprioception as the sixth major sense of the body, distinct from touch, which provided the nervous system with feedback about the changing position of the muscles and joints – information that is crucial for the automated control of posture and movement.

Sherrington was also interested in elucidating the neuroanatomical basis of more complex actions such as walking or running. These behaviours, he suspected, were the result of integrated and sequenced chains of much simpler reflexes within the spinal cord, which did not necessarily require the intervention of higher brain centres. To investigate this possibility, Sherrington examined the behaviour of decerebrate animals in which he made a complete transaction across the brain stem. In effect, it was an operation that produced an isolated length of spinal cord, with a good blood supply free of anaesthesia, in which he could analyse reflex pathways at will. Sherrington had first undertaken this procedure in 1896 and found it caused many muscles to become rigid and extended – a state he called "decerebrate rigidity". Sherrington went on to show that this muscle rigidity occurred because the higher forebrain areas normally inhibited the areas below. With this inhibition "shut off" the muscles then went into spasm from the "unchecked" excited activity of their sensory fibres, which stimulated the motor nerves innervating the muscles.

Despite the effects of muscular rigidity, Sherrington found he could induce a number of complex behaviours in his decerebrate animals. One resembled walking. For example, when Sherrington stimulated the left foreleg of a decerebrate cat, it stepped forward, while the corresponding left back leg moved backwards. And as this action occurred, the two legs on the right side of the body exhibited opposing movements. Clearly, these basic actions were mediated entirely by the spinal cord. Another spinal behaviour was the scratch reflex that Sherrington examined by inserting a small pin (dubbed an "artificial electrical flea") under the skin to simulate an itch over different parts of the body. Careful observation showed that this reflex was composed of two parts: the scratching motion and an adjustment for maintaining balance. The scratching motion consisted of a series of alternating knee, ankle and hip movements involving a total of nineteen muscles, while the animal utilised a further seventeen muscles to stop falling over. Remarkably, the decerebrate animal was able to follow the flea all over its body and scratch in the appropriate place.

Sherrington's thorough analysis of the reflex during the 1890s had convinced him that the neural circuitry underlying reflexive action consisted of individual nerve cells that were connected but not joined (i.e. fused). In this regard, he was a strong supporter of Cajal and cell theory and not the reticularist position taken by Golgi. In particular, Sherrington was impressed with Cajal's law of dynamic polarisation, which held that information flow through neurons was unidirectional – a finding he repeatedly confirmed in his experiments on spinal reflexes. Sherrington would also have been aware of the work of von Helmholtz in the 1850s, that had measured the speed of the nerve impulse – which was significantly faster than the reaction of his multi-component reflexes. And another factor that must have prompted Sherrington to reject the notion of fused neural pathways was the phenomena of neural inhibition. In 1897, when invited to write a chapter in a textbook on physiology (see also the previous chapter), Sherrington named this point of contact as the *synapse*. It would play an increasingly important part in his theorising on the nervous system.

In 1904, Sherrington was invited to give a series of ten Silliman lectures at Yale University. Two years later, these lectures were published as *The Integrative Action of the Nervous System*. Based on nearly two decades of research, the book attempted to provide a modern synthesis of nervous system function, which utilised concepts from reflex action and those borrowed from the fledgling neuron doctrine. But perhaps the greatest strength of *Integrative Action* was that it emphasised the importance of synapses. Although many still considered the synapse as a hypothetical entity,

Sherrington's meticulous examination of reflex activity had shown it to exist beyond all reasonable doubt. Moreover, it had important functional implications. For example, Sherrington had identified the synapse as the site where excitation and inhibition were exercised, which allowed reciprocal responses of different muscles to occur. In addition, Sherrington realised that the cell body of each motor neuron, whose axons ultimately provides the "final common pathway" to the muscles, must also receive a mixture of excitatory and inhibitory synaptic input from many diverse sources such as other areas of the spinal cord and higher brain areas. It was also apparent to Sherrington that all this information was being "summated" by the cell and formed the fundamental mechanism by which they make their decisions.

But perhaps the most important aspect of *Integrative Action* was that it provided a set of concepts by which the functioning of the nervous system could be understood, using the spinal reflex as the simplest unit of nervous co-ordination. Despite this, Sherrington realised the idea of a simple reflex was "a convenient, if not probable fiction" since all parts of the nervous system are intricately connected. In extending this idea, Sherrington explained how more complex behaviours arise from larger chains of simpler reflexes, often requiring the involvement of neural circuits at higher levels of the spinal cord and brain. Sherrington also recognised that as one ascends through the nervous system, the role of inhibition becomes more important and the complexity of reflexes and behaviour increases. Consequently, the most complex reflexes derive from the cerebral cortex, which Sherrington likened to a switchboard with millions of on-off switches, not unlike a great telephone exchange. Our behaviour, he argued, is therefore a product of vast numbers of nerve cells and synapses acting together in a beautifully organised and integrative way. Now recognised as one of the founding books of modern neuroscience, *Integrative Action* has been compared in greatness with Newton's *Principia Mathematica* and Harvey's *de Moto Cordi* (Levine 2007).

Sherrington moved to Oxford in 1913, where he became Waynflete professor of physiology until his retirement in 1936. By then, his physiological writings had spanned nearly 70 years and resulted in over 300 publications. Sherrington was also bestowed with many honours including the presidency of the Royal Society (1920–1925), a knighthood in 1922 and the Nobel Prize in 1932. Still continuing to write into his old age, Sherrington turned his attention to other subjects including philosophy and poetry. He also contemplated the relationship between mind and brain in *Man and his Nature* (1940) when he was 83 years old. Somewhat surprisingly, Sherrington admitted to believing in a form of Cartesian dualism in which the mind was not subject to physical laws – although it was somehow capable of interacting with, and directing, the brain's reflexive machinery. Famously, Sherrington also described the brain and mind as similar to an enchanted loom. Or as he put it: "The brain is waking and with it the mind is returning. It is as if the Milky Way entered upon some cosmic dance. Swiftly the head mass becomes an enchanted loom where millions of flashing shuttles weave a dissolving pattern . . . a shifting harmony of subpatterns". This idea has proven to be an enduring image and one inspiring countless students to contemplate how the mind arises from the electrical activity of the brain. Sherrington died from a sudden heart attack at the age of 94 in an Eastbourne nursing home.

Notes

1 This claim is made in Sherrington's Nobel Prize biography and in Liddell's (1952) biographical memoir of Sherrington, although it is strongly disputed in the book by Eccles and Gibson (page 4).

2 In the human body, there are thirty-one pairs of spinal nerves, one on each side of the vertebral column. Each has a dorsal and ventral root. The former carries sensory information into the spinal cord (and brain), whereas the latter convey motor input to the muscles.

Sherrington's major writings

Sherrington, C.S. (1906) *The Integrative Action of the Nervous System*. New York: Charles Scribner and Sons. Reprinted by Cambridge University Press, 1947 (with a new preface and full bibliography).

Sherrington, C.S. (1925) Remarks on some aspects of reflex inhibition. *Proceedings of the Royal Society B*, 97, 519–545.

Sherrington, C.S. (1933) *The Brain and its Mechanism*. Cambridge, UK: Cambridge University Press.

Sherrington, C.S. (1941) *Man on his Nature*. Cambridge, UK: Cambridge University Press.

Sherrington, C.S. (1965) Inhibition as a co-ordinative factor. In: *Nobel Lectures, Physiology or Medicine 1922–41*. Amsterdam: Elsevier.

Further reading

Burke, R.E. (2007) Sir Charles Sherrington's The Integrative Action of the Nervous System: A centenary appreciation. *Brain*, 130, 887–894.

Denny-Brown, D. (ed.) (1939) *Selected Writings of Sir Charles Sherrington*. London: Hamish Hamilton.

Eccles, J.C. and Gibson, W.C. (1979) *Sherrington: His Life and Thought*. New York: Springer International.

Granit, R. (1967) *Charles Scott Sherrington: An Appraisal*. New York: Double Day and Company.

Levine, D.N. (2007) Sherrington's "The Integrative Action of the Nervous System": A centennial appraisal. *Journal of the Neurological Sciences*, 253, 1–6.

Liddell, E.G.T. (1952) Charles Scott Sherrington 1857–1952. *Biographical Memoirs of Fellows of the Royal Society*, 8, 241–270.

Molnár, Z. and Brown, R.E. (2010) Insights into the life and work of Sir Charles Sherrington. *Nature Reviews Neuroscience*, 11, 429–435.

Swazey, J.P. (1969) *Reflexes and Motor Integration: Sherrington's Concept of Integrative Action*. Boston: Harvard University Press.

Chapter 3

John Newport Langley (1852–1925)

A Cambridge physiologist who more than any other person mapped out the functional neuroanatomy of the sympathetic and parasympathetic nervous systems (terms he coined in 1889) and who first proposed the concept of pharmacological receptors in 1905.

John Newport Langley was the son of a Newbury school teacher, who also received part of his formative education at Exeter Grammar School where his uncle was the headmaster. Entering St John's College, Cambridge, in 1871, Langley studied mathematics, history and literature, with the ambition of joining the Indian Civil Service. However, by his second year, he changed to natural sciences after coming under the influence of Michael Foster (later Sir) who became the university's first professor of physiology in 1883. Foster immediately recognised Langley's calibre and wrote "From the very first I marked him as a man of whom something was to be made". Indeed, Foster's estimation of Langley was vindicated when he was awarded first class in the Natural Science Tripos in 1874. A year later, Langley would graduate with a bachelor's degree in arts after studying biological sciences and take up the post of demonstrator. By then, he was actively involved in Foster's laboratory and together they wrote the popular *A Course of Elementary Practical Physiology and Histology*, first published in 1876 and which went through several editions. From then on, Langley did not deviate from his life's work in physiology and even eschewed studying for a medical degree as was customary in those days for researchers in physiology. In fact, apart from a few months in 1874 working in Germany, Langley would spend the rest of his life in Cambridge, a fellow of Trinity College and dedicated to research and teaching. He was elected a fellow of the Royal Society in 1883, took up dual posts as lecturer in natural science at Trinity College and University lecturer in physiology a year later and succeeded to the chair of physiology in 1903.

Even before his graduation, Langley had been given the task of examining the action of a South American plant extract called "jaborandi", which was known to contain an alkaloid[1] called "pilocarpine"[2]. This was just one of a number of alkaloids (others include nicotine, atropine and curare) that were attracting the attention of physiologists at the time. Langley soon discovered that an intravenous injection of pilocarpine in cats and dogs slowed down their heartbeat and caused marked salivation. The important question, however, concerned the reason why. There were two schools of thought: one believed that drugs such as pilocarpine acted directly on the nerve endings, while the other held they worked directly on the muscle or other bodily organs. Initially, Langley favoured the first explanation. He knew that the heart was innervated by the vagus nerve whose electrical stimulation led to a reduction in its beat. Thus, he assumed that pilocarpine was somehow acting on the same pathway. To test his theory, Langley applied curare to the vagus nerve – a procedure then widely believed to "paralyse" the nerve endings. If his theory was correct, then this should have inhibited the effects of pilocarpine. But this did not occur. Instead, the action of pilocarpine was only inhibited by the prior injection of another alkaloid called "atropine"[3] (derived from the *Atropa belladonna* plant). This was a puzzling effect and Langley could only explain it by assuming pilocarpine and atropine did not act on the vagus nerve after all but on the muscle of the heart itself.

In 1889, Langley, with the help of Dr William Lee Dickinson, turned their attention to the array of nerves that leave the length of the spinal cord and control the automated or vegetative functions of the body. At the time, this complex system of nerve pathways was known as the *involuntary* nervous system and they had been mapped by another Cambridge physiologist, Walter Holbrook Gaskell. In fact, Gaskell had found that these nerves left the spinal cord in three major outflows – corresponding to the spine's cranial, thoracic/lumbar and sacral regions. However, these three outflows had different anatomical characteristics. Although all were coated with a sheath of myelin as they left the spinal cord, only the thoracic/lumbar fibres entered the nearby sympathetic ganglia (these are a series of nerve cell clusters that run close to and down either side of the spinal cord). In contrast, the fibres of the cranial and sacral outflows passed further out into the body, where they terminated in ganglia closer to their target organs.

Gaskell then followed the nerve pathways beyond their ganglionic junctions to their final endings on the vital structures of the body – work showing that all these secondary fibres were unmyelinated. Thus, all the three major outflows of the involuntary nervous system had myelinated *preganglionic* and unmyelinated *postganglionic* fibres. Preliminary studies by Gaskell also showed that electrical stimulation of the sympathetic (thoracic/lumbar) outflow mimicked the effects of physiological arousal by exciting the organs they innervated. Hence, they speeded up the heart and increased respiration, etc. In marked contrast, stimulation of the cranial and sacral outflows had the opposite effect, leading to the inhibition of the very same organs.

It was against this backdrop that Langley and Dickinson made an unexpected discovery in 1889 with another alkaloid – nicotine.[4] Langley and Dickinson found that if a nicotine solution was applied directly to the ends of the preganglionic nerves in the sympathetic ganglion with a fine artist's brush (the nicotine easily penetrates into its cell bodies), then it blocked the transmission of nerve impulses from the preganglionic fibres of the thoracic/lumbar outflow to the postganglionic ones. So, for example, when Langley and Dickinson stimulated the preganglionic fibres of this system – a procedure that normally produced signs of physiological arousal – these effects were blocked by the nicotine application. While the pharmacological action of nicotine was as yet unknown,[5] this blocking effect proved to be a valuable tool for mapping out the functional and behavioural anatomy of this system. This work was also extended to the ganglion associated with cranial and sacral outflows – with similar results (i.e. the nicotine blocked the inhibitory effects of electrical stimulation of the preganglionic fibres).

By the end of the century, Langley had examined the specific physiological functions of the three major spinal outflows and begun to use a new nomenclature to describe this system of nerves. At first, he replaced the term "involuntary", which had been favoured by Gaskell, with "autonomic". But in 1905, he went further by describing the excitatory thoracic/lumbar system as "sympathetic" and the inhibitory cranial/sacral system as "parasympathetic". Today, these are standard terms that all students of neuroscience are (or certainly should be!) familiar with.

It is important to recap here what Langley had discovered. He had shown (1) that injections of pilocarpine slowed down the heart – an effect blocked by atropine and which appeared to be independent of the vagus nerve – and (2) that nicotine applied to the ganglia of both sympathetic and parasympathetic systems blocked the effect of preganglionic stimulation. Today, we can explain these effects by our pharmacological understanding of synapses. More specifically, we know that acetylcholine is the neurotransmitter liberated in the synapse between the preganglionic and postganglionic fibres in both the sympathetic and parasympathetic systems, and in the case of the vagus nerve, it is also the substance that is released by the parasympathetic postganglionic fibres, which slows down the heart. But at the turn of the century, these facts were not known. Although Sherrington had postulated the concept of synapses in 1897, there was still much uncertainty over their existence, and many physiologists, including Langley, were suspicious of chemical neurotransmission.

In 1901, Langley turned his attention to adrenaline, a substance produced by the adrenal glands, which exerted effects on the body very similar to those following electrical stimulation of the sympathetic nervous system. For example, adrenaline raised blood pressure, increased heartbeat, dilated the bronchial passages and stimulated salivary secretion. In an attempt to find out how adrenaline was producing these effects, Langley severed all the sympathetic nerve pathways and found that injections of adrenaline was still able to exert its physiological effects. Thus, adrenaline was acting on the target tissues directly. But the automatic effects of adrenaline were too rapid for the glandular source to be the origin – so where was the source of this chemical? One possibility was that adrenalin was being released from the sympathetic nerve endings. It is certain that Langley must have considered such a possibility, although he did not state it publicly. But a postgraduate student, Thomas Renton Elliott, working in Langley's laboratory at the time, was not so reticent. In 1904, when addressing the Physiological Society in London, Elliott proposed the existence of neurotransmission involving adrenaline. He also backed up his claims with evidence. Elliott pointed out, for example, that adrenaline only acted on tissue innervated by the *sympathetic* nerves and it had no effect on the parasympathetic system or skeletal muscle. And in concluding his talk, Elliott made the first clear modern statement about the feasibility of chemical neurotransmission. Or as he put it: "Adrenaline might then be the chemical stimulant liberated on each occasion when the impulse arrives at the periphery".

Most physiologists still remained sceptical about the merits of chemical neurotransmission – not least because electrical currents seemed to provide a much simpler and quicker mode of neurotransmission. Nor could Elliott explain how a chemical might produce a biological effect on its target tissue. However, in 1905, Langley proposed a simple answer to the later problem in a paper that attempted to explain the pharmacological actions of nicotine and curare. By now, Langley had extended his work with nicotine to the neuromuscular junction and found it also caused the skeletal muscles to contract. Furthermore, this effect occurred after the nerves to the muscles had been severed, indicating that nicotine was acting directly on the muscle. In addition, Langley found he could block nicotine's action on the contraction of the skeletal muscles by injecting the animal first with curare.[6] These findings were also reminiscent of the stimulatory effects of pilocarpine on the heart, which had been blocked by atropine.

To explain these effects, Langley proposed that nicotine must act on some specialised part of the muscle, which he termed its "receptive substance". In turn, the nicotine receptor then used the chemical message to initiate the biological effects inside the muscle cell. Langley also reasoned there must be two types of receptors: a nicotinic one that produces muscle contractions and another sensitive to curare that induced muscle paralysis.[7] Although Langley did not literally use the word "receptor", it was the first time this type of concept had been used to explain drug action. The receptor idea, however, was not entirely foreign as the term had been used by Paul Ehrlich in 1900 in an immunological context. Consequently, there were many biologists who recognised the feasibility of Langley's idea. In fact, the receptor theory is one of the key breakthroughs in the modern history of pharmacology, allowing a full understanding of drug action and supporting the theory of chemical neurotransmission. It was also a key moment in the history of neuroscience since it would be inconceivable today to understand the functioning of any type of nervous system without an understanding of synapses, receptors and neurotransmitters.

Langley may not have fully realised it during his life, but he had mapped out the sympathetic and parasympathetic nervous systems functionally and elucidated the chemical nature of their synapses.

Today, we know that the synapse between the preganglionic and postganglionic fibres of both these nervous systems is cholinergic with a nicotinic receptor. The synapse at the postganglionic endings of the sympathetic system is noradrenergic, while it is cholinergic (this time muscarinic) for the parasympathetic system. To complete the picture, the synapse at the neuromuscular junction is cholinergic (nicotinic). Although Langley was basically correct in believing that drugs react chemically with a receptor to initiate a biological effect, it took some time before the concept was widely accepted. For many, it was more plausible that the biological effects of a drug depended on its physical properties or the extent to which it could "dissolve" and penetrate the outer membrane of cells. In fact, the controversy over pharmacological receptors still remained unresolved at the time of Langley's death in 1925. Although evidence for the receptor theory would accumulate, especially through the work of Henry Dale, it remained largely hypothetical until receptor proteins began to be isolated from cell membranes in the 1970s. Only then was Langley's concept of receptor substances fully vindicated.

Notes

1 Alkaloids are a group of naturally occurring chemical compounds, which contain nitrogen atoms, made by plants and animals, which have a diverse range of physiological effects.
2 We now know that pilocarpine is an agonist at cholinergic muscarinic receptors.
3 We now know that atropine is an antagonist at cholinergic muscarinic receptors.
4 We now know that nicotine is a cholinergic agonist at nicotinic receptors.
5 Although nicotine is a stimulant, used in this experimental way it ultimately blocks synaptic transmission.
6 We now know that curare is an antagonist at cholinergic nicotinic receptors.
7 In this respect, we now know Langley was wrong since nicotine acts as an agonist, while curare is an antagonist, at the same receptor.

Langley's major writings

Langley, J.N. (1875) The action of jaborandi on the heart. *Journal of Anatomy and Physiology*, 10, 187–201.
Langley, J.N. (1903) The autonomic nervous system, Presidential Address. *Brain*, 26, 1–26.

Langley, J.N. (1905) On the reaction of cells and of nerve-endings to certain poisons, chiefly as regards the reaction of striated muscle to nicotine and to curari. *Journal of Physiology*, 33, 374–413.
Langley, J.N. (1906) Croonian Lecture, 1906 – On nerve endings and on special excitable substances in cells. *Proceedings of the Royal Society of London, Series B*, 78, 170–194.
Langley, J.N. (1921) *The Autonomic Nervous System. Part 1*. Cambridge, UK: Heffer and Sons (no second part published).

Further reading

Bennett, M.R. (2000) The concept of transmitter receptors. 100 years on. *Neuropharmacology*, 39, 525–546.
Davenport, H.W. (1991) Early history of the concept of chemical transmission of the nerve impulse. *The Physiologist*, 34, 129–190.
Fletcher, W.M. (1926) John Newport Langley in memoriam. *Journal of Physiology*, 61, 1–15.
Maehle, A.-H. (2004) "Receptor substances": John Newport Langley (1852–1925) and his path to a receptor theory of drug action. *Medical History*, 48, 153–174.
Nozdrachev, A.D. (2002) John Newport Langley and his construction of the autonomic (vegetative) nervous system. *Journal of Evolutionary Biochemistry and Physiology*, 38, 537–546.
Parascandola, J. and Jasensky, R. (1974) Origins of the receptor theory of drug action. *Bulletin of the History of Medicine*, 48, 199–220.
Prüll, C.-R., Maehle, A.-H. and Halliwell, R.F. (2009) *A Short History of the Drug Receptor Concept*. Basingstoke: Palgrave Macmillan.

Chapter 4

Korbinian Brodmann (1868–1918)

Brodmann was a German doctor who divided the human cerebrum into forty-three areas based on regional differences in the distribution, shape and size of its neurons – a mapping system still used today by researchers to guide them around the terrain of the cerebral cortex.

Korbinian Brodmann was born in Liggersdorf, Prussia (now southwest Germany) in 1868. The son of a farmer, Brodmann studied medicine in Munich, Berlin, and then Freiburg where he received his state licence to become a doctor in 1895. Soon after, Brodmann moved to Lausanne in Switzerland and then Munich where he worked treating psychiatric patients in a university clinic. Intent on setting up a private practice in the Black Forrest region, Brodmann's plans were thwarted when he contacted diphtheria, which led to a forced period of convalescence. In 1896, Brodmann took up a position as an assistant in a neurological clinic in Northern Bavaria under the directorship of the neurologist Oskar Vogt. It was an inspirational experience for the 28-year-old Brodmann who now decided to devote himself to the study of neurology and psychiatry. Described by Vogt as having broad scientific interests, a good gift of observation and great diligence in widening his knowledge, Brodmann then undertook his doctorate at the University of Leipzig where he wrote a thesis on ependymal sclerosis (a condition related to multiple sclerosis). Obtaining his MD in 1898, Brodmann worked in psychiatry at the University of Jenna and then at Frankfurt's Asylum for Lunatics and Epileptics from 1900 to 1901. Here, he met a number of researchers pioneering new techniques for staining nerve cells, including Carl Weigert, Franz Nissl and Alois Alzheimer – the later examining cortical degeneration in patients who had died from syphilis, which was then a major cause of mental illness. Indeed, it was Alzheimer who piqued Brodmann's interest in the cellular composition of the cerebral cortex – a problem that was to occupy him for the rest of his academic life.

In 1901, Brodmann accepted a position with Vogt who by now had established one of the largest private brain research institutes of its time in Berlin (the Neurobiologisches Institut) with excellent facilities for neuroanatomy. One of the scientists that Brodmann met here was the Pole Max Bielschowsky who had invented a silver impregnation stain that visualised degenerating nerve fibres. This stain would be put to good effect by Alzheimer in 1906, who discovered neurofibrillary tangles and plaques in the brain of a patient with a type of dementia that now bears his name. Vogt was also attempting to visualise nerve fibres in the brain, this time with a myelin stain, in order to create a topographic map of the main pathways in the cerebral cortex (a subject called "myeloarchitectonics"). As part of this project, Brodmann was encouraged to work with an alternative stain that had been developed by Nissl some years earlier. Known as cresyl violet staining, Nissl's technique was a cell body stain that highlighted the morphological shape of neurons and their main structural features.[1] Brodmann decided to use this technique to examine the cerebral cortical structure of a wide variety of mammals, including that of humans.

The first results of this endeavour appeared in 1903 when Brodmann and Vogt presented their work at the annual meeting of the German Psychiatric Society in Jena. Here, Brodmann described the different cytoarchitectonic structure of the precentral and postcentral gyri in the human cerebral cortex. Today, we know these areas as the motor cortex and somatosensory cortex, respectively – the former being located in the posterior frontal lobe and involved in the wilful elicitation of movement, while the latter lying adjacent to the motor cortex (although separated by the central sulcus) in the anterior parietal cortex, which receives touch and proprioceptive feedback from the body. Brodmann's work clearly revealed the structure, shape and position of the cells in these two regions to differ markedly – a finding supporting the idea that these two brain regions had very different behavioural functions.

This work also formed the basis of the first communication, of seven, which appeared in the *Journal für Psychologie und Neurologie* (of which Brodmann was editor) between 1903 and

1908. The sixth article published in 1908 is the most celebrated since it shows the first completed versions of the human cortical map in two parts – one figure showing the outer regions of the left hemisphere and the other exposing the medial side of the right. These two drawings display forty-seven distinct regions, now commonly known as Brodmann areas, and they have since become two of the most reproduced illustrations in the history of neuroscience. Brodmann's series of seven papers also formed the basis of his monograph published in 1909 entitled *Vergleichende Lokalisationslehre der Grosshirnrinde in ihren Prinzipien dargestellt auf Grund des Zellenbaues* (*The Principles of Comparative Localization in the Cerebral Cortex based on Cytoarchitectonics*). This book illustrated the cerebral cortex of some sixty-four different species, with reference to animals as diverse as the apes, lion, kangaroo and hedgehog – many of which Brodmann appears to have obtained from the zoo in Berlin. It was, in effect, an atlas on comparative localisation, which identified the homologies between the cortical areas of many different mammals through Brodmann's astute numbering system.

Again, as might be expected, the chapters illustrating the human cerebral cortex generated the most interest. Although his map of the human cortex on first sight displays forty-seven different cytoarchitectonic areas, with the numbers simply indicating the order in which he studied them, a close scrutiny of his diagrams actually shows that certain numbers (areas 12–16) are missing. Brodmann accounted for these "gaps" by explaining that some regions such as the olfactory, limbic and insular cortices are not identifiable in the human cortex but well developed in other mammalian species. To complicate matters, four other regions (numbers 48–52) are not shown in the human cortical map, although they appear elsewhere in the text. Nonetheless, it was apparent that the human cerebral cortex is significantly more complex than that found in monkeys and the great apes, which only have around thirty areas.

Although famed for its descriptive anatomy, Brodmann's book also makes several important contributions to cortical neuroscience. For example, at the time of its publication (1909), there was considerable confusion over the laminated structure of the cerebral cortex. While the cortex had been shown to have six layers by Rudolf Berlin in 1858, subsequent research had muddled the picture. Indeed, Brodmann notes that even Cajal had added to this confusion by postulating up to nine layers in man. Brodmann goes to considerable lengths, however, to make clear that the mammalian neocortex is always composed of six cellular layers – no matter what species is examined. Despite this, he shows that the human brain has thicker cortical layers along with increased cell size and greater density of neurons. Thus, the human cortex is cytoarchitectonically more complex than the cortex found in other primates.

Brodmann's views on cortical localisation is also discussed in his book. For the record, he described the theory of "psychic centres" as naïve since he believed higher mental functions were more likely to be neural processes that extended over the whole cortical surface and possibly involve subcortical regions as well. Nonetheless, Brodmann was willing to accept that individual regions did not "make equal physiological contributions to higher psychic processes". And, he also wrote that "the specific histological differentiation of cortical areas provides irrefutable proof of their specific functional differentiation". Brodmann was therefore advocating a compromise position in regards to localisation, whereby brain areas are specialised to some degree but nonetheless still process information in a holistic or distributed manner.

One might have imagined that Brodmann's monograph would have made him one of the most famous neuroanatomists in Germany. However, the institute where he worked was under financial pressure and Brodmann's position was neither permanent nor well paid. In an attempt to improve his circumstances, Brodmann submitted a thesis on the cerebral cortex of the lemurto the medical facility at the University of Berlin – work that should have resulted in him becoming a private doctor, which was a requisite for a tenured position. However, the thesis was rejected – presumably a decision more to do with the strained relations between the institute and the university than the quality of Brodmann's work. Brodmann relocated to Tübingen in 1910, where he worked as a physician and set up his own laboratory. In 1913, he became professor at the University of Tübingen, although the outbreak of World War I interrupted his research. After working in a hospital, in 1918, Brodmann finally managed to take up an appropriate post in the Psychiatric Research Institute at Munich, which was run by his friend Emil Kraepelin. During these last few years, Brodmann had been working on a second edition of his book. Sadly, he would never see it published. Brodmann developed pneumonia (possibly originating some years earlier from an autopsy) and succumbed to septicaemia just 5 days after falling ill.

Brodmann was not the first researcher to provide a cytoarchitectonic cortical map. This accolade goes to Australian Alfred Walter Campbell, working at Rainhill Asylum, Liverpool, who published a map outlining seventeen regions of the cortex in 1905. Nor was his the most detailed map of the times since Oskar Vogt and his wife Cecile would describe over 200 cortical areas on the basis of their myeloarchitectonics work in 1919. But Brodmann's topographic map fared the best and one can speculate why. Perhaps one reason lies with the two illustrations of the human brain Brodmann first published in 1906, which has captured the imagination of researchers and students ever since. Another reason may well be due to the number of brain areas delineated by Brodmann (i.e. around fifty), which seem to correspond reasonably well with the gross anatomical features of the cortex. But above all, Brodmann's parcellation of cortex areas has been shown in many cases to correlate well with different functions. Hence, Brodmann's areas 1, 2 and 3 correspond with the primary somatosensory area, area 4 is the motor cortex, area 17 is the primary visual cortex, and areas 44 and 45 overlap with Broca's area for language. Despite this, it is generally recognised that the correspondence between Brodmann's cytoarchitectonics and function is not always precise and should be viewed with caution. Yet, if anything, the use of Brodmann's areas has become even more popular today – especially with researchers concerned with investigating the cerebral cortex through the use of functional brain imaging techniques.

The fascination with mapping the cerebral cortex has never died. Indeed, one of the most remarkable attempts to extend Brodmann's topography was made by Constantin von Economo and Georg Koskinas at the University of Vienna in 1925. They produced a book and atlas of over 800 pages, containing 112 photographs and 162 figures, which identified 107 different cortical regions. Although this work has been described as a milestone in the history of science, it surprisingly never gained widespread acceptance. More recently, the Human Connectome Project, launched in 2009 by the National Institute of Health (NIH), whose goal is to build a complete "network map" or connectome of the human brain, fibre by fibre, has published results from highly detailed magnetic resonance imaging (MRI) scans of 201 healthy young adults. This has revealed a map showing 180 different regions – including ninety-seven "new" brain areas that have never been described before (Glasser et al. 2016). This map, it is argued, not only should help scientists working in the field of neuroimaging to be more certain about the areas of the brain they are observing, but should also assist many other types of brain scientist interested in structural-functional relationships. It remains to be seen whether this will replace Brodmann's parcellation of brain areas in the future.

The attempt to divide the cerebral cortex into discrete parts based on anatomical differences also raises important theoretical questions about the nature of brain function. Although it is apparent the division of cortex into different regions produces a bias that lends support to localisation of function, there are many who believe higher mental activity arises from brain activity taking place across neural networks spanning many different regions, rather than being confined to a single highly specialised region. Brodmann advocated the second view and it would appear that modern evidence from brain scanning techniques is beginning to support him – not least because many different psychological functions have been shown to recruit similar and overlapping brain regions and networks. Over a hundred years on from Brodmann's great work, neuroscience may finally be getting a step closer to reconciling the localisation versus holistic positions regarding the functions of the cerebral cortex.

Note

1 The Nissl stain works by making the rough endoplasmic reticulum within the neuron appear dark blue, due to the staining of ribosomal RNA.

Brodmann's major writings

Brodmann, K. (1909) *Vergleichende Lokalisationslehre der Grosshirnrinde in ihren Prinzipien dargestellt auf Grund des Zellenbaues*. Leipzig: Barth.

Garey, L.J. (1994) *Brodmann's Localization in the Cerebral Cortex* (English translation). London: Springer.

Further reading

Amunts, K. and Zilles, K. (2015) Architectonic mapping of the human brain beyond Brodmann. *Neuron*, 88, 1086–1107.

Draaisma, D. (2006) *Disturbances of the Mind*. Cambridge, UK: Cambridge University Press.

Glasser, M.F., Coalson, T.S., Robinson, E.C., Hacker, C.D., Harwell, J., Yacoub, E., Ugurbil, K, Andersson, J., Beckmann, C.F., Jenkinson, M.,

Smith, S.M. and Ven Essen, D.C. (2016) A multi-modal parcellation of human cerebral cortex. *Nature*, 536, 171–178.

Jones, E.G. (2008) Cortical maps and modern phrenology. *Brain*, 131, 2227–2233.

Judas, M., Cepanec, M. and Sedmak, G. (2012) Brodmann's map of the human cerebral cortex – or Brodmann's maps? *Translational Neuroscience*, 3, 67–74.

Loukas, M., Pennell, C., Groat, C., Tubbs, R.S. and Cohen-Gadol, A.A. (2010) Korbinian Brodmann (1868–1918) and his contribution to mapping the cerebral cortex. *Neurosurgery*, 68, 6–11.

Zilles, K. and Amunts, K. (2010) Centenary of Brodmann's map – Conception and fate. *Nature Reviews: Neuroscience*, 11, 139–145.

Chapter 5

Wilder Graves Penfield (1891–1976)

Penfield is best known for pioneering work on the surgical relief of epilepsy, which required the electrical stimulation of the brain in conscious patients – a procedure also enabling him to delineate the motor and somatosensory cortices, map language areas and explore facets of the human mind.

Wilder Penfield was born in the small town of Spokane, Washington, in 1891, a settlement of about a thousand people who were mostly miners and road builders. Penfield's father was a doctor and his mother a devout Christian, although their relationship was not a happy one and the couple would separate when Penfield was 8 years old – a parting that resulted in him moving with his mother in 1899 to Hudson, Wisconsin. Penfield once remarked that his only innate virtue was a "tenacity of purpose", and he would demonstrate this characteristic during his schooling when he overcome his "plodding mind" to graduate from high school as head of his class in 1909, thereby winning a place at Princeton. Here, Penfield studied English literature and excelled in sports, graduating with his bachelor's degree in 1913. It was also at Princeton that Wilder Penfield decided to follow the footsteps of his father and pursue medicine – because, as he once said "it seemed to be the best way to make the world a better place in which to live". However, from the age of 13, Penfield had harboured an ambition to win a Rhodes Scholarship enabling him to study at Oxford, and in 1914 after one failed attempt, he was successful. At Oxford, Penfield came under the influence of Charles Sherrington who inspired a lifelong fascination into the mysteries of brain and mind. With 2 years at Oxford, followed by a period of war service in France (he was wounded in 1916 when his ferry, the SS Sussex, was torpedoed), Penfield completed his medical degree at John Hopkins in 1918. The following year, Penfield moved to Boston as an apprentice to the famous brain surgeon Harvey Cushing. Returning to Oxford to complete the final year of his scholarship, Penfield then took up a research post at the National Hospital in London – a place where he began to develop his abiding interest in epilepsy. Now determined to learn more about this illness, Penfield practised surgery for 7 years at New York Presbyterian Hospital – a position allowing him to take up short-term research appointments elsewhere. This led him to work in Spain in 1924 under Pío del Río-Hortega who was famous for his discovery of microglia (cells that protect the brain from infection). Four years later, Penfield spent a few months in Germany where he learned how to remove brain scars to relieve focal epilepsy from Otfrid Foerster. During this time, Penfield also observed Foerster stimulating different parts of the cerebral cortex in fully conscious patients before his surgery – a procedure that Penfield would later make his own.

In 1928, at the age of 37, Penfield felt qualified enough to become a surgeon at the Royal Victoria Hospital in Montreal. He combined this with an academic role at nearby McGill University. A few months after his arrival, Penfield's sister Ruth was diagnosed with an advanced tumour deep in her brain – a growth that he attempted to remove with a radical operation. Unfortunately, he was unable to remove all the malignant cells and his sister died 3 years later. By now, Penfield's driving mission was to establish an institute where doctors, neurosurgeons and pathologists could work together in their pursuit of understanding and treating epilepsy. With an award of $1.2 million from the Rockefeller Foundation and the support of various other philanthropists including the government of Quebec, Penfield founded the Montreal Neurological Institute (MNI) in 1934. In the same year, Penfield became a Canadian citizen. Penfield would serve as the MNI's director until 1960, during which time he would see it become the world's premier centre for research into epilepsy.

Epilepsy afflicts about 1% of the population. Although in two-thirds of cases, a cause for the seizures cannot be found, there are other instances when the epileptic attacks are attributable to scar tissue in the brain, possibly arising through head injury, tumours or stroke. This scar tissue, which may interfere with the brain's electrical activity, can often be removed in an operation. However, one of the problems facing surgeons is not so much identifying the tissue but removing it in such a way that it does not cause any other secondary deficits. Surprisingly, much of the cerebral cortex can be removed without producing a behavioural impairment – although there are exceptions, especially when the lesion encroaches into areas responsible for movement and language. Thus, the surgeon must be highly mindful about whether their operation might lead to paralysis or speech dysfunction. To avoid this problem, Penfield, along with colleague Herbert Jasper, followed the lead of Otfrid Foerster and began electrically stimulating the exposed brain in fully awake patients prior to surgery. Requiring only a local anaesthetic applied to the scalp, this preliminary exploration enabled surgeons to remove tissue with greater confidence. The aim was simple: if the stimulation produced a sudden limb movement or caused the patient to vocalise, then this indicated a motor or language area to be spared. But in most instances, the patient was unresponsive to such stimulation, indicating that it could be removed safely.

Although Penfield was not the first to excise brain tissue for the relief of epilepsy (this was first undertaken by the nineteenth-century British doctor Sir Victor Horsley) or use electrical stimulation to investigate the brain, the Montreal procedure as it became known, in Penfield's hands, became a major advance in the surgical treatment of epilepsy. Not only were Penfield's rates of therapeutic success highly impressive with about two-thirds of patients showing improvement, but it was relatively safe with a mortality rate of less than 1%. Since its inception in the mid-1930s, the procedure has been improved and refined from research at the MNI and other centres, making it a standard practice across the world. In fact, over the course of his directorship, Penfield would perform over 2000 operations.

But of course, the stimulation procedure also gave fascinating insights into the functional organisation of the human brain, and the first major report of this appeared in 1937 when Penfield and Edwin Boldrey analysed the results from their first 163 operations. Their stimulation was mainly targeted at the deep fissure, called the "central sulcus", which separates the frontal and parietal lobes. The frontal part of this region, in dogs at least, had been found to contain isolated regions that produced movement by Gustave Fritsch and Eduard Hitzig in 1874. But, Penfield and Boldrey would show that this motor cortex was much more elaborate in humans. Not only was it confined within a strip of tissue called the "precentral gyrus", which lay in front of the central sulcus, but stimulation produced a highly organised set of bodily movements. Or to be more precise, the precentral gyrus contained an "upside-down" representation of the human body. Thus, stimulating the upper (superior) part of the gyrus produced feet movements, and more lateral and downward stimulation induced responses in the legs, abdomen and trunk, hands and head. It was also apparent that some parts of the body were more fully represented than others, with the face and hands having the largest amounts of tissue dedicated to them. To help illustrate this, Penfield and Boldrey drew a homunculus or "little man" to caricature the relative extent of each region.

Another brain region to be examined was the postcentral gyrus in the parietal lobes. Otherwise known as the somatosensory cortex, this area receives sensory, tactile and proprioceptive feedback from the skin and muscles of the body. Penfield and Boldrey mapped this region after patients reported a tingling or tactile sensation from a body region following stimulation. Again, it was found that this strip-like region was topographically organised with a map of the body resembling, although not identical, with the one in the motor cortex. For example, the somatosensory cortex had a greater proportion of tissue dedicated to input from the hands, lips and tongue that are capable of finer tactile discriminations. The topography of the motor and somatosensory cortices was famously illustrated in two drawings by Penfield and Rasmussen in 1950.

Language was another function that interested Penfield. It had been known since the mid-nineteenth century that a speech region (Broca's area) lay in the posterior frontal lobes, while a second one (Wernicke's area) was situated in the temporal lobes. Penfield localised the function of speech by getting his patients to read aloud and establishing whether any arrest of vocalisation resulted from brain stimulation. Although he confirmed that the area for speech production corresponded to Broca's area in the left frontal cerebral hemisphere, Penfield also realised that another frontal region, the supplementary motor cortex that lay just anterior to the precentral gyrus, also had an important language capacity. These findings showed that language was anatomically more complex than had been supposed. Moreover, Penfield also came to realise that no two patients had an exact correspondence of language areas in the brain – unlike the regions for movement and touch, which were predictable. This made it even more important for them to be carefully mapped out prior to surgery.

During the 1930s and 1940s, the main targets of Penfield's excisions were the frontal and parietal cortex – but over time he turned his interest to seizures originating in the temporal lobes. This type of epilepsy, first described by John Hughlings Jackson in 1876, is different to generalised seizures where the person turns rigid and falls to the ground with violent shaking. Instead, temporal lobe epilepsy, which is often pre-empted by a strange aura, fear or feeling of déjà vu, typically leads to a changed conscious state or "absence" with automated motor movements such as playing with the buttons of one's shirt. It is also not unusual for the person to become disoriented and wander away from their immediate situation. Despite this, temporal lobe epilepsy was rarely found to be caused by brain damage and thus not operated upon. In 1931, however, Penfield was stimulating the left temporal lobe of a woman with focal epilepsy when she suddenly saw herself re-enacting giving birth to her baby daughter – an event that had taken place years previously. Penfield gave little thought to this odd occurrence until something similar happened again some 5 years later. This time it was a 14-year-old girl who experienced herself walking through a meadow, followed by a man holding a snake. This memory was not only verified as genuine by the girl's mother, but one that often provided a warning of her daughter's impending seizure. This was a significant finding for Penfield because it implied that the site of the memory within the brain was close to the one producing the epilepsy. In other words, it was a suitable place for excision.

By the early 1960s, Penfield had distinguished two types of "psychical" response produced by his stimulation on the temporal lobes. The first were similar to the aura or déjà vu experiences that preceded the start of the seizure, and Penfield referred to them as *interpretive* since they were essentially a disturbance (or illusion) of the patient's interpretation of their current situation. They also often gave rise to largely irrational emotions such as fear. The second type of response, however, appeared to be memories from the past, which Penfield referred to as *experiential* since the patient was re-experiencing the event in some way. Although some described these as dream-like or resembling a flashback from the past, all patients agreed they were more vivid than could be recalled voluntarily. Penfield also found that such episodes followed a real time sequence of events, which moved forward in time until the stimulating electrode was withdrawn.

Penfield drew several conclusions from these remarkable observations. First, he believed that experiential responses were real memories from the patients past, and for the most part they recalled unimportant moments in the patient's life, suggesting to Penfield that nothing is lost from the memory record. Second, he came to view memory as "a record of the stream of consciousness" analogous to a strip of cinematographic film on which all our experiences are stored. Today, both these are views are not shared by the vast majority of cognitive psychologists or neuroscientists. A more probable explanation is that experiential responses and memories are a reconstruction and not an exact record. It is also interesting to note that Penfield was well aware that he was never able to localise the human mind to any cortical area – a position supported by the fact that his patients were always conscious of their immediate surroundings when reporting their elicited memories in the operating room.

By the time Penfield retired in 1960, at the age of 70, his life's work had been awarded with numerous honours including the British Order of Merit in 1953 and the accolade of being the greatest living Canadian in 1956 despite being American by birth.[1] Many other awards would be bestowed on Penfield in his retirement, which he spent increasingly concerned with philosophical issues in books such as *The Mystery of the Mind* (1974), written in his eighties. Penfield still had some surprises for his readers, admitting that as a young man he had been inspired by the material belief that the neurological processes of the brain would one day explain the mind. However, by the end of his life, he had changed his opinion. As Penfield pointed out, the conscious willing to do, or believe in something, had never been invoked in a person through electrical stimulation of their brain. Or as he put it in a memorable analogy "There is a switchboard operator as well as a switchboard". In fact, Penfield had come to believe that the brain and mind are two entirely different things – a dualist view most famously associated with Descartes. Penfield died at the age of 85, just 3 weeks after completing his autobiography *No Man Alone* – a book espousing his belief in team work as a means of improving surgery and understanding the mysteries of the brain.

Note

1 This accolade was made by Writer Eric Hutton in the Canadian magazine *Maclean's*.

Penfield's major writings

Jasper, H. and Penfield, W. (1954) *Epilepsy and the Functional Anatomy of the Human Brain*. Boston: Little Brown and Co.
Penfield, W. and Rasmussen, T. (1950) *The Cerebral Cortex of Man*. New York: Macmillan.
Penfield, W. (1955) The twenty-ninth Maudsley lecture: The role of the temporal cortex in certain psychical phenomena. *The Journal of Mental Science*, 101, 451–465.
Penfield, W. (1958) *The Excitable Cortex in Conscious Man*. Liverpool: Liverpool University Press.
Penfield, W. and Roberts, L. (1959) *Speech and Brain Mechanisms*. Princeton: Princeton University Press.
Penfield, W. (1967) Epilepsy: The great teacher. *Archives of Neurology, Scandinavia*, 43, 1–10.
Penfield, W. (1975) *The Mystery of the Mind. A Critical Study of Consciousness and the Human Brain*. Princeton: Princeton University Press.
Penfield, W. (1977) *No Man Alone. A Neurosurgeon's Life*. Boston: Little Brown and Co.

Further reading

Feindel, W. (1977) Wilder Penfield: His legacy to neurology. *Canadian Medical Association Journal*, 116, 1365–1377.
Feindel, W. (1982) The contribution of Wilder Penfield to the functional anatomy of the human brain. *Human Neurobiology*, 1, 231–234.
Jefferson, L. (1981) *Something Hidden: A Biography of Wilder Penfield*. Toronto: Doubleday and Co.
Rasmussen, T. (1976) Wilder Penfield, 1891–1976. *Journal of Neurosurgery*, 45, 248–250.
Ruelland, J.G. (2005) Wilder G. Penfield (1891–1976), neurosurgeon and scientist. *Vesalius: Acta Internationales Historiae Medicinae*, 11, 64–69.
Snyder, P.J. and Whitaker, H.A. (2013) Neurologic heuristics and artistic whimsy: The cerebral cartography of Wilder Penfield. *Journal of the History of the Neurosciences*, 22, 277–291.
Whitaker, H.A. and Ojemann, G.A. (1995) From Bartholow to Penfield: The historical development of electrical stimulation of the human brain from 1874 to 1928. *Brain & Language*, 51, 60–62.

Chapter 6

Edgar Douglas Adrian (1889–1977)

Adrian was a prolific researcher who was the first to record electrical activity from single axons that established frequency coding in an all-or-nothing manner as the means by which neurons transmit information.

Edgar Douglas Adrian was born in Hampstead, London, and was the son of eminent barrister Alfred Douglas Adrian who was a legal advisor to the British Local Government Board – a supervisory body overseeing local administration in England and Wales. His mother was the daughter of noted mathematician Charles Howard Barton. In 1903, Adrian entered Westminster School as a King's scholar and 5 years later enrolled at Trinity College, Cambridge, where he studied a range of subjects that made up the Natural Sciences Tripos degree. He passed the first part of this degree in 1910 with first class marks and followed this success by going on to specialise in physiology for his final year, obtaining a BA degree with first class honours in 1911. During this time, Adrian would come to the attention of the distinguished Cambridge physiologist Keith Lucas whose laboratory he joined in 1912. A year later at the age of 24, Adrian was elected to a fellowship of Trinity College on account of his research prowess. Despite this, he continued working towards his medical degree and spent the war years treating soldiers with nerve injuries at St Bartholomew's Hospital in London. He completed his medical degree in 1915 and after the war returned to Cambridge where he devoted his life to teaching and research.

At this particular time, in physiology, there was a major debate concerning the nature of information that passed down the nerve axon. The issue was whether the strength of the nerve impulse could vary in intensity (i.e. amplitude) or if it always occurred in full, or simply not at all. In 1904, Lucas had electrically stimulated the frog's cutaneous dorsi muscle (a small muscle from which he could cut strips containing some twelve to thirty muscle fibres) and showed that as he increased the stimulation more fibres were bought into action in a step-by-step manner. This provided evidence for the all-or-nothing principle, although doubts still remained over the issue. For example, it was known that the electrical strength of a nerve impulse weakened if it passed through an anaesthetised section of nerve fibre – a finding implying a partial response might still be possible. Adrian was given the task of examining this effect, and to do so he exposed a small segment of frog nerve to alcohol vapour and tracked the impulse along the fibre. The results showed the alcohol block only produced a short-lasting decrement in the signal's force and the nerve impulse was fully reinstated as it moved away from the blockade. Thus, providing the signal was not totally extinguished, it would be recharged back to its full strength. This and a series of similar experiments between 1912 and 1914 made a compelling case for the all-or-nothing law – research that led to Adrian winning the Cambridge Washington Gold Medal for physiology.

Adrian's research was put on hold by World War I, which also claimed the life of Lucas, who was killed in a mid-air collision while testing equipment for the Royal Air Force in 1916. Consequently, it was left to Adrian to edit and complete Lucas's posthumous work *The Conduction of the Nervous Impulse* in 1917. Returning to Cambridge in 1919, Adrian took over Lucas's laboratory, where he continued his investigations into the nature of the nerve impulse. By the early 1920s, practically all the work in this field had been done on motor nerves because the recording instruments of the time, such as the galvanometer and capillary electrometer, were not sensitive enough to detect the minute impulses produced by stimulation of the sensory nerves. But this situation changed after the war when a number of technological advances had taken place, enabling researchers to record electrical signals with greater fidelity. One of these improvements was the thermionic valve (or vacuum tube), which had been adapted by the intelligence services to amplify and intercept coded messages. However, it could also be used to record signals from nerve fibres, and the first person to do this was a friend of Adrian, the Harvard professor Alexander Forbes, who built the first vacuum-tube amplifier for neurophysiological work in 1918. Adrian quickly realised the importance of the new device, which could amplify a tiny nervous impulse some fifty times without distortion, and he invited Forbes to Cambridge to help him construct his own. But this was only the beginning. Further refinements in amplification over the next decade would enable Adrian to magnify the strength of neural signals many thousands of times more, and other inventions such as the cathode ray oscilloscope allowed visual plots of rapid voltage fluctuations of nerve impulses to be displayed on film. It would lead to an era that some have dubbed the heroic age of neurophysiology.

During the early 1920s, Adrian and Forbes used the amplifying powers of the thermionic valve to prove that the all-or-nothing principle also applied to sensory nerves – in this case, the saphenous nerve of the cat. But this work soon led to further developments. For example, in another study, Adrian recorded activity from nerve fibres that were still attached to the gastrocnemius muscle of the frog's leg. Adrian found that if he hung weights from the muscle, this caused the number of impulses in the nerve fibre to increase from 10 impulses per second at rest to 50 impulses when fully stretched. It was apparent that the firing rate was encoding the amount of force being placed on the muscle – the first time nerve impulses had been shown to be encoding sensory information. But this effect was short-lasting: if the force remained constant, the nerve impulses began diminishing in frequency after a few minutes. Similarly, if Adrian pinched the skin, he obtained a quick succession of action currents that soon decreased as he maintained the pressure. This type of adaptation is known as "habituation" and is a form of sensory learning that allows the brain to ignore information that is no longer significant.

Over the next few years, along with various collaborators, Adrian showed that the all-or-none principle and adaptation were universal across different types of nerve and various species. These were exciting times for Adrian, and in a letter to Forbes he wrote that he was having such a fine time with his new amplifier that he could barely bring himself to take a holiday. But Adrian was also preparing the ground for one of the milestones of nerve physiology research by discovering a way of recording electrical activity from single neurons. Up to this point, all research on the electrophysiological properties of nerve cells had been performed on peripheral fibres that were actually bundles of much finer individual axons. This meant every recorded nervous event was a composite response – often involving hundreds of nerve axons firing in unison. Although this had given rise to the all-or-none principle theory, along with the first demonstration of frequency coding, these laws still had to be confirmed at the level of the single nerve cell. But this was seemingly impossible. Researchers had no way of being able to dissect out a tiny single axon, embedded and entwined among a morass containing hundreds of others within a given fibre.

In 1925, however, Adrian began working with a young Swedish researcher called Yngve Zotterman who had joined his laboratory on a Rockefeller Fellowship. One of the tasks the two men set themselves was to isolate the sensory receptor embedded within muscle tissue that encodes stretch – a long, slender, elongated set of fibres known as the muscle spindle, whose significance had been first recognised by Charles Sherrington in the 1890s. The first attempt to do this using the frog's stenocutaneous muscle which was known to contain only a few spindles was unsuccessful – that is until Zotterman hit upon the idea of progressively shaving away fine slices of muscle, until enough tissue was stripped away to isolate a slither that contained a single spindle. It was an important breakthrough because the spindle was attached to a single nerve axon. And it was a straightforward task to record from this single axon with an electrode, even though it lay in a bundle containing a mass of inactive cells. By using a vacuum tube to amplify the impulse and passing it through a capillary electrometer that projected the signal onto a photographic screen, Adrian and Zotterman were the first to record and visualise the electrical activity from individual nerve fibres. This appeared as a wave form "like a beating heart" with a regular frequency and amplitude. As Zotterman later remarked, they had discovered "a great secret of life".

When Adrian used an oscilloscope to visualise the impulses from the single axon, they appeared as small rapidly occurring spikes on the screen. This also showed that the action potential was always of the same height and travelled at the same speed regardless of the stimulus that had caused it. Thus, the all-or-none principle applied to the transmission of information by single neurons. Interestingly, Adrian's results also demonstrated that a single axon could transmit up to 400 nerve impulses per second. This high figure was not totally unexpected, as it was known that the action potential only lasted a few milliseconds (msecs) (i.e. thousandths of a second). Clearly, the axon could convey a lot of information in a short period of time. But it was the patterning of the responses which was the important factor. As Adrian stretched the spindle, the number of spikes per second increased, with the number of impulses proportional to the strength of the sensory stimulus (or its change). It was almost as if the nerve cells were encoding information with a simple form of Morse code. In fact, Adrian used a memorable analogy in 1927 when he likened the coded nervous message to a stream of bullets fired from a machine gun. Indeed, if the nervous impulses are fed through an amplifier and played through a loudspeaker, this is exactly what they sound like.

Adrian and Zotterman extended their single unit recordings to other types of cell including sensory neurons transmitting touch information from skin of the cat paw, along with other receptors involved in pain. A few years later in 1929, Adrian collaborated with the American Detlev Bronk to examine motor neurons again, including the phrenic nerve controlling the diaphragm. No matter what type of axon he examined, they always encoded information in the same way with small and brief electrical impulses of the same intensity. In 1932, Adrian was confident enough to proclaim that the all-or-nothing law was a "universal" coding mechanism used by every type of nerve cell.

In 1932, at the age of 43, Adrian was awarded the Nobel Prize along with Sir Charles Sherrington "for their discoveries regarding the functions of neurons". By now, Adrian was increasingly turning his interest to rhythmical nervous activity, especially in the brain. For example, he had begun recording from the respiratory centre of the isolated goldfish brain and demonstrating that it had a beating neural rhythm of some 20 to 60 impulses a minute, which corresponded to the gill movements of an intact fish. Adrian also became interested in the work of the German Hans Berger who had invented the electroencephalograph (EEG) in 1924 – an apparatus that used scalp electrodes to record and amplify the tiny voltages of the brain. Recognising that this could be a useful experimental device for examining the brain of humans, Adrian not only helped bring Berger's work to much wider recognition but also extended his findings in a number of significant ways. Adrian's interest in the EEG would also do much to establish EEG as a clinical speciality, including its adoption as a diagnostic tool in epilepsy.

Adrian also undertook pioneering work on the cerebellum. For example, in the early 1940s, he showed that different regions of the cerebellum responded with electrical discharge activity after stimulation of the cerebral motor cortex. The cerebellar activity was replicated when various parts of the body were moved, including the forelimb, hind limb and head. This work was also extended to other areas of the brain including the parietal lobes, which led Adrian to discover a secondary area of the somatosensory cortex beyond the postcentral gyrus, which is now recognised as being important in sensorimotor integration with involvement in attention, learning and memory. Towards the end of his career, Adrian also advanced our understanding of smell, especially by identifying the contributions of sensory nose receptors in stimulating the mitral units of the olfactory bulb.

Sadly, Adrian's research career came to an abrupt end in 1958 when water leaked into his laboratory and destroyed all his equipment. He was now in his late sixties and unwilling to start afresh with new apparatus. Nonetheless, he continued lecturing and acted successfully as the chancellor of the University of Cambridge from 1967 to 1975. Adrian had also been knighted Baron of Cambridge in 1955 – one of the last hereditary titles granted in England. It was a position enabling Adrian to sit in the House of Lords where he spoke on such matters as scientific policy, nuclear disarmament and biological weapons. Because his son Richard produced no heir, however, the title became extinct in 1995. A man with many active interests outside his work including painting and walking in the Lake District, Adrian also apparently (according to the historian Stanley Finger) had a fondness for zooming around Cambridge at breakneck speed on a motorcycle with a side car. The author of several important books, Adrian had numerous honours bestowed upon him, including the presidency of the Royal Society (1950–1955) and the Royal Society of Medicine (1960–1962). He also held honorary degrees from over thirty universities and various memberships and fellowships of learned societies too numerous to mention here. Following his wife's death in 1965, Adrian moved back into Trinity College where he lived until his health began to decline. He spent his final 2 weeks in a nursing home where he died in 1977 at the age of 87.

Adrian's major writings

Adrian, E.D. (1926) The impulses produced by sensory nerve endings. *Journal of Physiology*, 61, 49–72.

Adrian, E.D. and Zotterman, Y. (1926) The impulses produced by sensory nerve endings, part 2. *Journal of Physiology*, 61, 151–171.

Adrian, E.D. (1928) *The Basis of Sensation*. New York: Norton and Co.
Adrian, E.D. (1932) The activity of the nerve fibres. In: *Nobel Lectures: Physiology or Medicine 1922–1941*. Amsterdam: Elsevier.
Adrian, E.D. (1932) *The Mechanism of Nervous Action: Electrical Studies of the Neurone*. Philadelphia: University of Philadelphia Press.
Adrian, E.D. (1946) *The Physical Background of Perception*. Oxford: Clarendon Press.
Adrian, E.D. (1954) Memorable experiences in research. *Diabetes*, 3, 17–18.

Further reading

Bradley, J.K. and Tansey, E.M. (1996) The coming of the electronic age to the Cambridge Physiological Laboratory: E.D. Adrian's valve amplifier in 1921. *Notes and Records of the Royal Society of London*, 50, 217–228.
Finger, S. (2000) *Minds Behind the Brains: A History of the Pioneers and Their Discoveries*. Oxford: Oxford University Press.
Frank, R.G. (1994) Instruments, nerve action, and the all-or-none principle. *Osiris*, 9, 208–235.
Garson, J. (2015) The birth of information in the brain: Edgar Adrian and the vacuum tube. *Science in Context*, 28, 31–52.
Hodgkin, A.L. (1979) Edgar Douglas Adrian, Baron Adrian of Cambridge. *Biographical Memoirs of the Royal Society*, 25, 1–73.
McComas, A.J. (2011) *Galvani's Spark: The Story of the Nerve Impulse*. Oxford: Oxford University Press.
Zotterman, Y. (1964) Neurophysiology in the nineteen twenties. *The Rockefeller Institute Review* 2, 4, 1–4.

Chapter 7

Henry Hallet Dale (1875–1968)

If one person can be given the accolade of greatest British pharmacologist of the twentieth century, then a strong case can be made for Henry Dale who not only identified acetylcholine as a transmitter at several sites in the nervous system but also postulated the existence of nicotinic and muscarinic receptors.

Dale was born in London in 1875 into a relatively prosperous middle-class family. After being educated at small public schools, Dale won a scholarship to the Leys School in Cambridge in 1891 – a Methodist institution strongly focussed on getting its students into the colleges of its local university. The school encouraged scientific pursuits, and it was here that Dale made his first forays into biological experimentation with frog muscle-nerve preparations and simple histological techniques. After winning a scholarship to enter Trinity College in 1894 to study natural sciences, Dale came under the influence of Walter Gaskell and John Langley leading to his first scientific paper in 1900. In the same year, Dale left Cambridge to finish his medical training at St Bartholomew's Hospital, London, which was accomplished in 1903. Dale then began a yearlong research studentship at University College. In his semi-autobiographical *Adventures in Physiology*, Dale claims he was faced with a rather bleak academic prospect at this point and decided to accept an appointment at the newly built Wellcome Physiological Research laboratories in London. This was an unusual choice since the Wellcome Institute was a non-profiting concern, created by the American Henry S. Wellcome (later Sir), aimed at scientific advancement for the benefit of all. In fact, this objective had already been met when the early Wellcome laboratories produced a diphtheria antitoxin in 1890. Although some of Dale's university friends warned him that it would limit his academic independence, the Wellcome offered a secure income and career advancement. Dale was to take full advantage of this situation by becoming its director in 1906.

When Dale started work in his new job, he was encouraged by Henry Wellcome to examine the chemical constituents of a parasitic fungus *Claviceps purpurea*, otherwise known as ergot. This had been known to infect cereals such as rye in medieval times, leading to epidemics of St Anthony's fire characterised by convulsions, hallucinations and intense burning sensations in its victims. Later, in the eighteenth century, ergot was found to cause constriction of the blood vessels, resulting in its adoption by midwives to induce abortions or stop bleeding after childbirth. Dale admitted he was "not at all attracted by the prospect" of making his first excursion into pharmacology by examining "the ergot morass". Yet, Wellcome's suggestion turned out to be a good one. As Dale began to examine ergot, he realised it was a treasure trove of active pharmacological substances, many of which were isolated by his colleague George Barger.

Dale's first breakthrough with ergot came through a stroke of good fortune. One day he was asked to test an extract from the adrenal glands to see if it contained adrenaline. He tested the extract on a cat pre-treated with ergot and found the animal's adrenal-induced blood pressure did not increase as expected. Thus, there was something in the ergot inhibiting the effects of the extract. This turned out to be a substance called "ergotoxine", which blocked the constricting effects of adrenaline on the smooth muscle of blood vessels. Dale had not only discovered the first alpha-adrenergic blocking agents, now routinely used to treat a range of medical conditions, but also guessed that receptors for adrenaline must exist on the arterioles, which he called "adrenoreceptors". This discovery also made Dale examine the role of noradrenaline (NA) more closely – a closely related substance that was commercially available. He soon realised that NA mimicked the effects of sympathetic nervous stimulation more faithfully than adrenaline itself. Although Dale did not go as far to suggest that NA was acting as a neurotransmitter in the sympathetic system, this would be shown to be the case by other pharmacologists in the 1940s.

In 1910, Dale identified another chemical in ergot, which caused smooth muscle to constrict. But more interestingly, injections of this chemical in animals produced many symptoms of anaphylactic shock – a potentially life-threatening response associated with bad allergic reactions. In fact, Dale had discovered histamine, shown a year later to be a naturally occurring chemical after he isolated it from the walls of the intestine. Over the next decade, Dale showed that histamine is found in many other bodily regions and released in response to injury. In turn, histamine then caused constriction of the arteries along with a general "leakage" of the capillaries – a reaction supporting the immune response. These discoveries would prove of great importance in surgery, especially during World War I, when the control of anaphylactic shock after battle injuries helped avoid serious complications. Later, in the 1940s, histamine was found to exist in the brain and finally identified as a neurotransmitter in the 1960s.

Dale's most famous contribution to pharmacology, however, was his elucidation of the role of acetylcholine in autonomic and somatic nervous function. This story began in 1913 when Dale was given a routine extract of ergot to test. He injected it into an anaesthetised cat and expected to see a modest rise in arterial blood pressure – although on this occasion it had the opposite effect. At first, Dale thought he had inadvertently injected an air bubble into the animal, but with repeated testing he realised it was a real pharmacological effect, leading him to suspect that ergot might contain an unsuspected chemical known as "muscarine". This chemical, first extracted from the iconic red and white spotted fly agaric mushroom (*Amanita muscaria*), was of great interest to researchers at the time since it produced effects that closely mimicked the action of the parasympathetic nervous system. After working with colleague Arthur Ewins, however, Dale realised the depressor substance in ergot was not muscarine but acetylcholine. The implication, although Dale was careful not to endorse it at the time, was that acetylcholine acted as a neurotransmitter in the parasympathetic system – a controversial idea, not least because it had yet to be found in the body.

Although muscarine and acetylcholine had similar physiological effects, there were some notable exceptions. One was at the neuromuscular junction where only acetylcholine caused muscle contraction. And this site showed another interesting characteristic: the effects of acetylcholine were mimicked by nicotine and both were blocked by curare. In contrast, the action of muscarine on the parasympathetic system was blocked by the alkaloid atropine. Dale offered an explanation for these results in 1914 when he proposed there were two types of cholinergic receptors with different chemical affinities. He called these *nicotinic* and *muscarinic*. This still did not prove the concept of chemical neurotransmission at the synapse, although it was an important step in its development.

In 1914, Dale left his post at the Wellcome to join the Medical Research Committee where he became head of the Department of Biochemistry and Pharmacology. Dale's work on acetylcholine, however, would be put on hold due to the outbreak of World War I. Now, Dale turned his focus to examining traumatic wound shock, which he had already shown was linked to the release of histamine. This soon led Dale to find an effective remedy by using a plasma transfusion to increase the circulating blood volume. Dale also undertook tests on drugs produced by British and French manufacturers to replace German pharmaceuticals – work that impressed upon him the need for much more stringent regulation. Many of his recommendations for improvements in drug standardisation and safety would become commonplace and adopted internationally.

Although the possibility of chemical neurotransmission had first been raised by Emil du Bois Reymond in 1877 and the concept of chemical receptors given credence in 1905 by John Langley, both ideas were contentious. Many researchers believed that synapses were points of contact between nerves, with communication between them more likely to be electrical than chemical. Dale, for one, was not prepared to risk his reputation by speculating on the possibility of chemical neurotransmission when he only had tentative evidence. More definite proof was needed and this was provided by a friend of Dale's – the German Otto Loewi in 1920, who undertook an experiment that has become one of the most famous in pharmacology. Loewi claimed that on the night of Easter Saturday, 1920, he awoke from a dream with an idea for an experimental procedure. After scribbling down his thoughts, Loewi returned to sleep, only to find he could not decipher his notes the next morning. The following night, Loewi dreamt the same dream, and this time he cycled straight to his laboratory to perform the experiment. The procedure was simple. Loewi took a heart from a frog still attached to its parasympathetic (vagus) nerve and placed it into a chamber of saline. He then stimulated the vagus nerve that caused the heartbeat to slow. Then Loewi removed some of the solution from the chamber and applied it to a new heart whose vagus nerve had been removed. This now caused the second heart also began to slow down. The most likely explanation was that the first heart had secreted a chemical into the saline, which then acted on the second heart.

Loewi suspected that the chemical being liberated by the first heart was acetylcholine, and while it would take 5 years and a large number of experiments to prove, he was ultimately shown to be correct. Indeed, Loewi would add to the evidence by finding small traces of cholinesterase (an enzyme that breaks down acetylcholine) in heart tissue in 1930. He also showed that atropine blocked the effects of the solution taken from the donor heart – a finding implicating the muscarinic receptor.

Before acetylcholine could be confirmed as a neurotransmitter, however, it needed to be identified as a naturally occurring chemical in the body. This was no easy task. If acetylcholine was being released from nerve endings, then the amounts were minuscule and most probably broken down by cholinesterase. In fact, Dale was confronting much the same difficulty with histamine. But the doubts over histamine were dispelled in 1927 when Dale and Harold Dudley managed to isolate it from the liver and lungs. During this work, Dale also discovered potent cholinergic activity in extracts of horse spleen. Although Dale had no idea why spleen contained acetylcholine, he set about visiting a local slaughter house and collecting tissue from freshly killed horses and oxen. From this endeavour, in 1929, Dale and Dudley obtained 71 pounds of spleen tissue, from which they extracted one-third of a gram of acetylcholine. It was a tiny amount but enough to prove acetylcholine was a natural constituent of biological tissue.

By the early 1930s, most pharmacologists accepted that the parasympathetic branch of the vagus nerve secreted acetylcholine. But there was little evidence to show it acted in a similar way elsewhere. This changed in 1933 when Dale was able to employ Wilhelm Feldberg, who had been forced to leave the University of Berlin after the Nazi party came to power. Feldberg had developed an effective way of detecting cholinergic activity that involved recording the contractions of a leech muscle which had been pre-treated with the cholinesterase inhibitor eserine (physostigmine). Remarkably, this increased the muscle's sensitivity to acetylcholine more than a million fold. And it was a technique sensitive enough to detect the presence of acetylcholine extracted from the nerve endings of animals such as cats and dogs.

To do this, Feldberg stimulated a given nerve in an animal and collected blood drawn from a nearby vein – hoping it would reveal the tiny amounts of acetylcholine when it was applied to the leech muscle. This method proved so successful that it pinpointed the sites of acetylcholine release throughout the peripheral nervous systems. It showed that acetylcholine was not only secreted by the endings of the parasympathetic nervous system but also at the synapse between the preganglionic and postganglionic fibres of the sympathetic system. Adding to this, acetylcholine was also confirmed as the neurotransmitter at the neuromuscular junction of skeletal muscles. In 1936, Dale and Loewi shared the Nobel Prize "for their discoveries relating to chemical transmission of nerve impulses". It has been described as the greatest advance in neuroscience of the twentieth century. Even so, not all physiologists were won over by the concept of chemical neurotransmitters, and the issue would remain debateable until the mid-1950s, largely as a result of work undertaken by one of Dale's adversaries – John Eccles.

Dale remained at the National Institute for Medical Research (NIMR) until his retirement in 1942. By then, he had won numerous awards including a knighthood in 1932 and the Royal Society's Copley medal in 1937 (the society's highest honour). This was followed by the Order of the British Empire in 1948. Dale also served as president of the Royal Society from 1940 to 1945 and president of the Royal Society of Medicine from 1948 to 1950. A man who was said to be affable and unstinting in his support of colleagues, although at times austere, Dale was nonetheless regarded in enormous affection by all who knew him. After retiring, Dale continued working on government committees and was a trustee for the Wellcome Trust, overseeing grant awards for research until his eighty-fifth birthday in 1960. Even then, he only cut back his work a little, remaining as a scientific advisor until 1 week before he died in a Cambridge nursing home at the age of 93 in 1968. Dale is widely regarded as the founder of biochemical pharmacology, and there are few who have made such a colossal impact on their respective fields. And, remarkably, he is the only key thinker in this book who never held an academic post at an university.

Dale's major writings

Dale, H.H. (1933) Progress in autopharmacology. *Bulletin of the John Hopkins Hospital*, 53, 297–329.

Dale, H.H. (1948) Accident and opportunism in medical research. *British Medical Journal*, 2, 451–455.
Dale, H.H. (1953) *Adventures in Physiology*. London: The Wellcome Trust.
Dale, H.H. (1954) *An Autumn Gleaming: Occasional Lectures and Addresses*. London: Pergamon Press.
Dale, H.H. (1954) The beginnings and prospects of neurohumoral transmission. *Pharmacological Reviews*, 6, 7–13.
Dale, H.H. (1958) Autobiographical Sketch. *Perspectives in Biology and Medicine*, 1, 125–137.

Further reading

Davenport, H.W. (1991) Early history of the concept of chemical transmission of the nerve impulse. *The Physiologist*, 34, 129–190.
Feldberg, W. (1969) Henry Hallett Dale 1875–1968. *British Journal of Pharmacology*, 35, 1–9.
Fick, G.R. (1987) Henry Dale's involvement in the verification and acceptance of the theory of neurochemical transmission: A lady in hiding. *The Journal of the History of Medicine and Allied Sciences*, 42, 467–485.
Fishman, M.C. (1972) Sir Henry Hallett Dale and the acetylcholine story. *Yale Journal of Biology and Medicine*, 45, 104–118.
Koelle, G.B. (1986) Henry Hallett Dale 1875–1968. *Trends in Pharmacological Sciences*, 7, 288–289.
Schild, H.O. (1976) Dale and the development of pharmacology. *British Journal of Pharmacology*, 56, 3–7.
Tansey, E.M. (2003) Henry Dale, histamine and anaphylaxis: reflections on the role of chance in the history of allergy. *Studies in History and Philosophy of Biological and Biomedical Sciences*, 34, 455–472.
Tansey, E.M. (2006) Henry Dale and the discovery of acetylcholine. *Comtes Rendus Biologies*, 329, 419–425.

Chapter 8

Walter Rudolf Hess (1881–1973)

Hess was a Swiss physiologist who pioneered the use of electrical brain stimulation in unanaesthetised freely moving animals – work that showed the importance of the diencephalon (especially the hypothalamus) in the regulation of the sympathetic and parasympathetic nervous systems.

Walter Rudolf Hess was born in the Swiss town of Frauenfeld that lies about 40 km north-east of Zurich close to the German border. The son of a college physics professor who had his own purpose built laboratory at home, Hess was encouraged to take an interest in science from an early age. But it was the family doctor when treating the young Hess for tuberculosis who had the bigger influence by evoking his interest in medicine. After 7 years of gymnasium education, Hess began his medical studies at the age of 19 at the University of Lausanne. As was customary for medical training at the time, Hess studied at various universities including Bern and Zurich in Switzerland, along with Kiel and Berlin in Germany. During one of his early dissection classes, Hess came across an anomaly in the arteries of the foot, which piqued his interest. Not satisfied with the answer provided by his teacher, Hess arrived at his own conclusion about the cause of the abnormality, which he felt had been produced by the blood's mechanical forces. It would lead to his first paper in 1903.

Hess obtained his medical degree from the University of Zurich in 1906 with a thesis on the dynamics of blood flow. After a short spell as an assistant surgeon in a state hospital, Hess decided his professional interest lay in ophthalmology that promised more financial security. It would also appear Hess was well suited to his new vocation since he invented an apparatus, called the "Hess screen", able to measure eye paralysis in diplopia, which is still in use today. In 1908, Hess took over an ophthalmologic practice in the medieval town of Rapperswill on Lake Zurich, which allowed him to marry and start a family. But it seems that Hess was never entirely happy with his profession and yearned for a career in physiological research. In 1912, Hess was given that chance when offered an assistant's position in the Physiological Institute at the University of Zurich. Although this entailed some financial risk, Hess took the opportunity. It proved to be a wise decision for just 4 years later, in 1917, Hess was promoted to chair of physiology – a remarkable achievement since the position had a long line of eminent predecessors and normally only awarded to Germans. Hess was to serve in this position until his retirement in 1951.

At first, Hess continued his investigations into the haemodynamics of the blood, although his work was interrupted by World War I when he was called upon to serve as captain in the Swiss Army Medical Corps. Despite these restrictions, Hess managed to spend a year of his war time (1915/1916) at the University of Bonn, Germany, in the laboratory of Max Verworn who was one of the leading physiologists in Europe. At the time, Verworn was undertaking important research into the all-or-none principle of axonal conduction and had also used electrical stimulation as a means of investigating the nervous system – a technique that Hess would later use to great effect. After the war, Hess turned his attention to the higher reflex control of arteries and heart, which inevitably led him to become interested in the autonomic nervous system – the array of nerves that control the vital organs of the body. This would also lead Hess to visit Cambridge where he met John Langley – who had named, mapped out and elucidated the functions of the sympathetic and parasympathetic systems (see Chapter 3).

By the mid-1920s, Hess had turned his attention to the brain's control (or in his terminology, the central operational elements) of the autonomic nervous system. His first paper on the subject appeared in 1925 entitled *On the interrelations between psychic and vegetative functions* (in German), and it was apparent he had recognised the importance of a brain area with direct links to the spinal cord, which lay directly under the cerebrum called the diencephalon. This region contained the hypothalamus and thalamus, and was known largely through the work of Walter Cannon and Philip Bard at Harvard that it exerted an important influence on the visceral actions of the body, including those involved in emotion. Hess's first paper also shows he was giving considerable thought to a research paradigm that would shape his research for many years to come – namely, the localised stimulation and ablation of these same brain areas.

Although electrical stimulation of the brain was not new, having been used in 1870 to discover the motor cortex in dogs by Gustav Fritsch and Eduard Hitzig in Germany (who used galvanic stimulation) and then David Ferrier in England who used faradic stimulation on monkeys, in the hands of Hess the technique would become much more sophisticated. To begin, he used the smallest possible stainless steel electrodes and established the minimal current that was effective at the lowest intensities (using low-frequency direct current pulses). Consequently, the stimulation had little spread to other brain areas. Hess also came to pay a great deal of attention to the precise guiding of the electrodes and the localisation of stimulation points, and for this purpose he developed his own atlas based on a series of complete histological sections in three cardinal planes of the cat brain. This allowed Hess to map the anatomically complex diencephalon millimetre by millimetre. In fact, Hess usually tested each of his placements with at least three different stimulating potentials and film the results, thereby producing a huge amount of data that Hess collated on card files. As he noted in the introduction to his *The Functional Organization of the Diencephalon* (1957), this system enabled him to find his way round tens of thousands of records and to arrive at unequivocal conclusions based on the comparison of different animals and stimulation points.

Hess also used lesioning to destroy the same areas he had been stimulating. This he undertook by using the same electrodes, except in this case they were connected to an apparatus that generated a high-frequency alternating current, which heated and coagulated the tissue at the electrode tip. By using this method, Hess was hoping to identify deficits in behaviour, which would give functional clues to the destroyed part of the brain. It therefore provided a second way of assessing the autonomic functions of the diencephalon. Later, in the 1930s, Hess would use the Marchi method of staining the degenerating nerve fibres from his lesioning to map out of the neural pathways underlying his stimulation effects.

The first preliminary reports of Hess's experimental procedure appeared in 1927 and they concerned simple responses to stimulation, such as heart rate, respiration and blood pressure. Of particular interest, however, was his discovery that sleep could be induced in cats by stimulating an area that included the medial thalamus. This also produced a synchronisation of EEG patterns in the cerebral cortex characteristic of sleep. These were significant findings and Hess presented them to an International Congress in Boston in 1929. Later, Hess would show that sleep could also be induced by stimulating the anterior regions of the hypothalamus.

But Hess's most famous work would take place in the 1940s and focus on the ventral and posterior areas of the cat's hypothalamus. Providing the right intensity of current was used, electrical stimulation of these regions immediately turned a good natured cat into a bad tempered one – with hissing, growling, spiting, tail lashing and the bristling of hair along its back. It was a type of response seen when a cat is confronted by a dangerous threat, and one inevitably causing the animal to flee from the situation or initiating its own attack in defence. Even the experimenters were not immune from the aggression, for if they strayed to close, they would often have the cat lurch and viciously claw at their legs. Not surprisingly, therefore, Hess's laboratory assistants carried a towel at these moments in order to protect themselves. Hess was also able to evoke many of the individual components of this behaviour (piloerection, unsheathing of claws, etc.) by stimulating exact points in the hypothalamus. But almost as dramatic was the effect when the stimulation was turned off – in this instance, the cat quickly returned to its normal calm state. Lesions to the same areas generally produced a loss of attention, reduced muscle tone and lack of spontaneous movement, often resulting in the cat becoming listless and not concerned with normal cleanliness.

It was clear to all observers that the physiological arousal elicited in the cat by the hypothalamic stimulation was activating the sympathetic nervous system and preparing the animal for the so-called flight-or-fight response. Looked at from another perspective, the sympathetic system could also be understood as part of the nervous system designed to expend energy. Consequently, Hess referred to the ventral and posterior regions of the hypothalamus as the "ergotropic zone" (ergo is derived from the Greek word meaning "work"). Its prime purpose, according to Hess, was to produce behaviours that assist the organism to greater success in its stressful conflicts with its environment.

As Hess placed his electrodes more anteriorly through the hypothalamus, however, a different pattern of behaviours emerged. Now, stimulation led to a reduction in physiological arousal with a fall in blood pressure, heart rate and respiration, accompanied by increased gastrointestinal mobility and salivation. Another reaction was panting – a response normally seen in the dog when it is hot, and one Hess believed was equivalent to sweating in man. With higher levels of stimulation, Hess also observed signs of hunger and thirst, cleaning behaviours, urination and defecation. But low-level stimulation of the anterior hypothalamus typically induced a marked relaxation of behaviour, which after a few minutes often caused the cat to curl up and fall asleep. This effect was further observed after stimulating the lateral part of the massa intermedia – a band of gray matter close to the medial third ventricle that connects both sides of the thalamus. Hess also showed that this type of sleep was highly naturalistic for, if a piece of meat was held in front of the cat's nose, it would react to the smell, awaken and seize the food, before lying down again and continuing its rest.

Hess recognised that many of these responses were characteristic of those governed by the parasympathetic nervous system, which acts to reduce physiological arousal while also governing the body's organs and glands during rest. The parasympathetic system is also operative during digestion, and this led Hess to designate the anterior hypothalamus as the "trophotropic zone" (tropho is derived from the Greek word meaning "nourishment"). But it was this nervous system's control of rest activity that was most intriguing – for it indicated that sleep had a vegetative regulatory influence that needed to be carried out in conditions of calm when energetic activity is reduced to a minimum. Viewed in this way, sleep is more than just simple rest – it is active recuperation – a radical idea for the 1920s and 1930s.

Over the course of his career, Hess undertook research on over 400 animals and examined at least 3500 areas of stimulation. Thus, a huge body of anatomical and behavioural data involving the hypothalamus, thalamus and adjacent regions of the midbrain were analysed and interpreted. This relentless experimental pursuit was also performed by a small research team composed of a couple of research assistants, a mechanic, a secretary and a graduate student. Nonetheless, their endeavours would result in over 300 research papers and a number of monographs. And it would also lead to Hess being awarded the Nobel Prize (shared with Egas Moniz) in

1949 "for the discovery of the functional organisation of the diencephalon as a coordinating centre of visceral function". At the time, Hess was 68 years old and still actively working. When he retired 2 years later in 1951, he had made Zurich into a globally renowned centre of brain research.

Despite his retirement, Hess became professor emeritus and borrowed an office from a colleague, where he continued to evaluate his experimental findings. As is often the case in retired neuroscientists, Hess became increasingly interested in the mind-brain problem, which he wrote about in the 1960s. For him, psychological functions were an essential feature of the living organism, especially in humans where conscious experience represented the supreme level of behavioural integration. But Hess also believed that subjective experience and voluntary action was not scientifically intelligible. Indeed, he doubted whether we would ever find an explanation of how a material brain, using a code of neural signals, could transform itself into a human mind with all its psychic experiences, feelings and willful impulses. In 1967, Hess moved to Ascona in southern Switzerland, known for its Mediterranean-style architecture and cafe-lined promenade on Lake Maggiore. According to his granddaughter who paid him weekly visits, Hess was still driven by scientific curiosity and bred snakes and dissected crabs in his apartment, while in the garden he experimented with exotic plants and grew various species of grapes, pears, apples and figs. She also recalls her grandfather had a love of pets and in his younger years was accompanied almost everywhere, including the laboratory, by his faithful dog. In later years, his favourite pet was a talking gray parrot called "Joko", whom he had bought up from a young bird. Joko was always by his side or perched on his shoulder. Hess died of heart failure in 1973 at the age of 92 in Locarno, Switzerland.

Hess's major writings

Hess, W.R. (1957) *The Functional Organization of the Diencephalon*. New York: Grune and Stratton. Edited by John R. Hughes.

Hess, W.R. (1963) From medical practice to theoretical medicine. An autobiographical sketch. *Perspectives in Biology and Medicine*, 6, 400–423.

Hess, W.R. (1964) *The Biology of the Mind*. Chicago: University of Chicago Press.

Hess, W.R. (1964) The central control of the activity of internal organs. In: *Nobel Lectures, Physiology or Medicine 1942–1962*. Amsterdam: Elsevier.

Hess, W.R. (1967) Causality, consciousness and cerebral organization. *Science*, 158, 1297–1283.

Further reading

Akert, K. (1981) *Biologic Order and Brain Organization: Selected Works of W.R. Hess*. Berlin, Heidelberg; New York: Springer-Verlag.

Akert, K. (1999) Walter Rudolf Hess (1881–1973) and his contribution to neuroscience. *Journal of the History of the Neurosciences*, 8, 248–263.

Hess, C.W. (2008) Walter R. Hess (17.3.1881–12.8.1973). *Schweizer Archives Fur Neurologie and Psyciatrie*, 159, 255–261.

Jung, R. (1981) Walter R. Hess (1881–1973). *Review of Physiology, Biochemistry and Pharmacology*, 88, 1–21.

MacDonald, D.A. (1951) W.R. Hess: The control of the autonomic nervous system by the hypothalamus. *The Lancet*, 257 (6655), 627–629.

Chapter 9

Karl Spencer Lashley (1890–1958)

Lashley is famous for his attempts to localise the site of memory storage in the brain – a research endeavour that took over three decades and led him to conclude that the engram is stored throughout the cerebral cortex with all its regions playing an equal role in storage.

Karl Spencer Lashley was born in the small town of Davis, West Virginia. His father was a sometime store keeper, postmaster and banker in the town, who also served as mayor for several terms. His mother was a school teacher with literacy and artistic interests, and it appears she had the bigger influence on Lashley's development who recalled he was reared on a combination of Nestlé's food and Nietzsche. Although most of Lashley's childhood was spent in Davis, there was a 4-year period when the family moved around a great deal and spent part of it joining the gold rush to the Klondike in north-western Canada. This experience made a lasting impact on the young boy who later retraced his steps as an adult. Despite these interruptions, Lashley graduated from high school at the young age of 14 and then enrolled at the University of West Virginia where he intended to study Latin but was persuaded to change to English by an admissions tutor. A year later, Lashley changed to zoology. After his Bachelor of Arts in 1910, Lashley won a teaching fellowship at the University of Pittsburgh where he taught and wrote a master's thesis on the bacteriology of rotten eggs. A year later, Lashley started his doctorate at John Hopkins in Baltimore. Although intending to follow zoology, Lashley came under the influence of psychologist John Watson in whose laboratory he increasingly spent time. This led to his first publication in 1912 on visual discrimination in rats, with several other psychological papers following, including a contribution to his mentor's book *Behavior* in 1914 – the same year he completed his PhD (in zoology). In fact, Watson would deliver the presidential address to the American Psychological Association in 1915, which was the first account of his conditioning research (on the salivary reflex) and acknowledge Lashley's importance in its development.

During the academic year of 1915, Lashley continued to work with Watson on a Johnston Scholarship where he investigated the effects of strychnine and other drugs on maze learning in rats. Lashley would also publish observations made during a field trip on the nesting activities of terns and colour vision in birds. And it was also around this time that Lashley decided to make psychology his chosen career. But the close association with Watson was not to last. By now, Watson was developing a greater interest in infant behaviour and Lashley more inclined to take a biological approach to conditioning research, which he pursued by making contact with the surgeon Shepherd Ivory Franz in Washington, who was noted for his work on brain-injured patients. Franz also carried out lesion experiments on rats and monkeys to back up his clinical work, and this provided a good opportunity for Lashley to learn new skills, including the surgical ablation of the cortex. This led to Lashley's first published papers on cerebral lesioning in rats with Franz in 1917.

The next few years saw Lashley change academic direction once more. In 1917, he accepted an instructorship in psychology at the University of Minnesota, although it appears that conditions were not ideal, and a year later Lashley began a post with the United States International Hygiene board in Washington where he was given the job of educating the public with respect to the dangers of venereal disease – a task involving the presentation of motion films, which he jointly shared with Watson. In 1920, Lashley returned to the University of Minnesota as assistant professorship, rising to full professor in 1924. During this time, he followed several interests including the neurobiological basis of learning. However, it would be his move to Chicago in 1926 that saw Lashley undertake some of his most famous work on the neural locus of the conditioned reflex in the cerebral cortex. This was a period where psychologists were strongly influenced by the work of the Russian Ivan Pavlov who believed all learning took place through a process of *classical conditioning* in which natural or unconditioned stimuli could become paired with neutral stimuli to form new conditioned or learned responses. Lashley had introduced the work of Pavlov to

Watson in 1915 and the notion of the conditioned reflex soon underpinned his theory of behaviourism, which held that all behaviour was ultimately the result of similar reflexive stimulus-response contingencies. It was a doctrine that was to dominate psychology up until the 1950s – and one that Lashley would eventually come to reject.

Pavlov had not only gone to great lengths to describe the behavioural processes of classical conditioning but hypothesised that such learning must come about through the creation of new neural connections in the brain. Or more precisely, he supposed that conditioning occurs when a new neural pathway is created that links a cortical sensory region with one mediating the behavioural response. A useful analogy is one of an elaborate and complex telephone switchboard in which the operator makes a new connection. Lashley was strongly influenced by this theory since he recognised if this was the case, then lesions to the cerebral cortex should disrupt, or abolish, the learning and retention of conditioned responses. In addition, by using the lesion method, it should be possible to trace the conditioned reflex pathways through the cortex and identify the site of memory storage (or the *engram*).

Lashley trained rats on problems of varying difficulty, including puzzle boxes, visual discrimination tasks and mazes. He then assessed whether knife cuts, or lesions, made to the cortex prior to training, or those made after learning, disrupted performance. However, the results of his extensive experimentation, which culminated in the monograph *Brain Mechanisms and Intelligence* in 1929, were to some readers highly surprising. Put simply, Lashley showed that the anatomical site of the lesion was unimportant, although its size did have a bearing on the animal's performance or ability to re-learn the task. For example, after training his rats to run through an eight alley maze, Lashley found no impairment in retention when a small lesion was made anywhere in the cortex. Thus, no cortical site was more critical than any other for maze learning. In fact, Lashley could remove up to a fifth of the cortex without producing a significant effect on performance. With larger lesions, deficits did begin to show, although Lashley had to destroy nearly half the cortex to reveal a significant loss of learning – and even then the rat could re-learn the task reasonably quickly with further training.

On the basis of these findings, Lashley derived two general laws about the cortical basis of learning and memory: (1) learning is mediated by the whole cerebral cortex and stored in neural circuits that are widely and diffusely distributed – a principle he called *mass action* and (2) all parts of the cerebral cortex play an equal role in the storage of memory (which in effect means all of its regions are equally able to perform a task) – a principle he called *equipotentiality*. These two laws implied that the neural correlates of learning were stored everywhere or no place in particular. Moreover, they not only refuted Pavlov's stimulus-response theory of cortical function but also dealt a serious blow to the assumptions of behaviourism with its simple emphasis on explaining behaviour in terms of conditioned reflexes.

The conclusions of *Brain Mechanisms and Intelligence* continue to reverberate in modern day neuroscience for they relate to a central issue in our attempts to understand the brain – namely, to what extent are its functions localised in discrete areas, or diffusely spread as an integrated whole. The debate has a long history. In the eighteenth century, Franz Gall was the first to propose localisation of cortical functioning when he developed phrenology, although his theories were soon discredited by Pierre Flourens who (like Lashley) undertook lesioning studies on a wide range of animals and concluded that "The faculty of sensation, perception and volition is essentially . . . one faculty". During the latter part of the nineteenth century, however, the pendulum began to swing back to the localisation camp, with Paul Broca discovering a language centre (which now bears his name) in the left frontal cortex in 1861 and Gustave Fritsch and Eduard Hitzig identifying the motor cortex in 1870. Both these findings were later supported by Wilder Penfield in his studies involving electrical stimulation of the human brain. Nonetheless, there are many who have disputed the concept of localisation, including Hughlings Jackson, Korbinian Brodmann and more recently Karl Pribram.

Although Lashley appears to have convincingly refuted localisation theory, there are some grounds to dispute his claims. It can be argued, for example, that Lashley's use of visual discrimination and maze learning tasks are too complex to be explained away as being stored in simple engrams. Hence, it may be that the problem of maze learning not only requires spatial learning (e.g. the hippocampus), along with tactile and proprioceptive learning (e.g. somatosensory cortex), but also involves other areas of the brain involved in vision, olfaction and emotional functioning – all of which are combined in performing the task. Thus, a partial lesion of the cerebral cortex may only disrupt part of this multi-modal learning, leaving the rat with adequate memory to negotiate the maze again. It can also be argued that while Lashley highlighted the great difficulty of locating engrams, his theory was unsatisfactory because he was unable to replace it with anything better. Indeed, in 1950, when reviewing much of his life's work, Lashley wryly expressed his own frustration at the situation when he wrote, "I sometimes feel in reviewing the evidence that the necessary conclusion is that learning just is not possible. It is difficult to conceive of a mechanism which can satisfy the conditions set for it".

In 1935, Lashley became professor of psychology at Harvard and, concurrently with the latter half of his tenure, he took up the directorship of the Yerkes Laboratories of Primate Biology in Orange Park, Florida. At the time, this institution housed the largest collection of chimpanzees for research purposes in the world and would attract many leading researchers. Lashley remained in these two positions until his retirement in 1955. Over these 20 years, Lashley undertook a wide variety of research. His interest in the visual system, for example, led him to study the neural connections between the lateral geniculate nucleus and the striate cortex – work revealing there was something approaching a point-to-point projection of the geniculate upon the visual cortex. Later, working with George Clark, Lashley provided evidence to show that the cerebral cortex only contained six architectonic regions, which was far fewer than that proposed by Brodmann. In another interesting set of experiments, Lashley tested the Gestalt theory of perception, otherwise known as "psychological isomorphism", which held that the neural representation of perceptual events are encoded as force fields in the brain that bear an actual structural resemblance to the visual events themselves. However, Lashley found no support for this theory when he disrupted these hypothetical force fields by covering the cerebrum with highly conductive gold foil or implanting chips of mica into the cortex.

But Lashley's most important contribution to psychology during his last research years took place in 1948 when he participated in the 5-day *Hixon Symposium on Cerebral Mechanisms in Behavior* held at the Californian Institute of Technology. This symposium is now recognised as providing a significant step in the cognitive psychology revolution that overthrew the prevailing doctrine of behaviourism and created a new agenda for research. Arguably its key speaker was Lashley who discussed the problem of serial order in behaviour. Lashley's main point was that many skilled acts, such as running a maze, playing a musical instrument, speaking words in a sentence or returning a high velocity serve at tennis, cannot be explained by a chain of simple conditioned reflexes in which one element of an action provides a stimulus for the next. For example, Lashley calculated the speed of finger movements involved in playing a rapid cadenza on a piano and compared this with the known speed of neural transmission. It revealed that the intervals between successive finger movements were far too brief to support the stimulus-response theory. In other words, there was not enough time for a sensory message from the finger to reach the motor areas of the brain and then return to the finger muscles. The only way this serial order could be achieved according to Lashley was through a central integrative process and not by a linkage of successive elements. In short, there must be a more complex representation in the brain, or "schemata of action", able to generate behavioural responses independent of any stimulus. This was a radically new conception of the brain and one totally at odds with the principles of behaviourism.

Lashley's last published article was entitled *Cerebral Organization and Behavior* in 1958, which dealt with "the problem of how the brain knows that it knows; what characteristics of neural activity constitute the mind". In this paper, Lashley leaves the reader in no doubt he was an avowed materialist who strongly believed that the human mind will one day be fully accountable by the operations of the brain. For him, there was no logical reason to doubt why a full identity between mind and brain could not be achieved. In February 1954, at the age of 67, while undertaking his annual two-weekly stint of teaching at Harvard, Lashley collapsed and was diagnosed with haemolytic anaemia. He was given a long-term course of cortisone treatment, although this sadly led to complications that necessitated a splenectomy. He was apparently well on the way to recovery when he died on holiday in Poitiers, France, with his second wife in 1958.

Lashley's major writings

Lashley, K.S. (1929) *Brain Mechanisms and Intelligence: A Quantitative Study of Injuries to the Brain*. Chicago: Chicago University Press.

Lashley, K.S. (1930) Basic neural mechanisms in behavior. *Psychological Review*, 37, 1–24.

Lashley, K.S. (1943) Studies of cerebral function in learning: XII. Loss of the maze habit after occipital lesions in blind rats. *Journal of Comparative Neurology*, 79, 431–462.
Lashley, K.S. (1950) In search of the engram. *Symposia of the Society of Experimental Biology*, 4, 454–482.
Lashley, K.S. (1951) The problem of serial order in behavior. In: L.A. Jeffress (ed.), *Cerebral Mechanisms in Behavior*. New York: John Wiley.
Lashley, K.S., Chow, K.L. and Semmes, J. (1951) An examination of the electrical field theory of cerebral integration. *Psychological Review*, 58, 123–136.
Lashley, K.S. (1958) Cerebral organization and behavior. *Research Publications of the Association of Research into Nervous and Mental Disease*, 36, 1–18.

Further reading

Bartlett, F.C. (1960) Karl Spencer Lashley, 1890–1958. *Biographical Memoirs of the Royal Society*, 5, 106–118.
Beach, F.A., Hebb, D.O., Morgan, C.T. and Nissen, H.W. (1960) *The Neuropsychology of Lashley*. New York: McGraw Hill.
Dewsbury, D.A. (2002) Constructing representations of Karl Spencer Lashley. *Journal of the History of the Behavioral Sciences*, 38, 225–245.
Dewsbury, D.A. (2005) *Monkey Farm: A History of the Yerkes Laboratories of Primate Biology, Orange Park, Florida, 1930–1965*. Lewisburg, PA: Bucknell University Press.
Hebb, D.O. (1959) Karl Spencer Lashley 1890–1958. *American Journal of Psychology*, 72, 142–149.
Orbach, J. (ed.) (1982) *Neuropsychology after Lashley: Fifty Years since the Publication of "Brain Mechanisms and Intelligence"*. Hillsdale, NJ: Erlbaum.
Weidman, N.M. (2008) *Constructing Scientific Psychology: Karl Lashley's Mind-Brain Debates*. Cambridge, UK: Cambridge University Press.

Chapter 10

Giuseppe Moruzzi (1910–1986) and Horace Winchell Magoun (1907–1991)

In 1951, Moruzzi and Magoun published a landmark paper, showing the existence of an ascending reticular formation that governed the states of sleep and waking though its control over the rhythmical electrical activity of the cerebral cortex.

Giuseppe Moruzzi was born in the small town of Campagnola Emilla, some 60 km north-west of Bologna in northern Italy, although grew up in Parma where his father was a physician. The family had a long tradition of medical practice with Moruzzi's great-grandfather Giovanni Cavacciuti being professor of general pathology at the University of Parma in the 1840s and his uncle a collaborator of the great French neurologist Jean-Marie Charcot. In 1927, Moruzzi began medicine at the University of Parma where he studied under the neuroanatomist Antonio Pensa who had been a pupil of Camillo Golgi. His interest in neuroanatomy at this stage was reflected in his first published research paper in 1930, examining the granular layer of the cerebellum. After graduating in 1933, Moruzzi worked as an assistant under the neurophysiologist Mario Camis whom he followed to the University of Bologna in 1936 after he was offered the chair of physiology.[1] A year later, Moruzzi spent a year at the University of Brussels in the laboratory of noted sleep researcher Frédéric Bremer. This was a period when the neurophysiology of sleep was poorly understood, although Bremer's theory that saw sleep as a passive process which occurred when sensory information was "turned off" from reaching the cerebral cortex was dominant. Little could Moruzzi have realised, some 15 years later, he would perform a seminal experiment (with Magoun) that strongly refuted this idea. In late 1938, Moruzzi joined the Nobel laureate Edgar Adrian in Cambridge on a Rockefeller

Foundation Fellowship, where he recorded the neural activity of the pyramidal tract (a pathway from the motor complex to spinal cord) in cats. Keen to continue this collaboration with Adrian, Moruzzi won a Lewis Fellowship in 1939, but the outbreak of World War II forced him back to Bologna where he spent several years serving in the medical corps of the Italian army. After the war, Moruzzi was appointed as professor of physiology at the University of Parma and then at the University of Pisa in 1948. These were economically difficult times in Italy following its loss in the war, and Moruzzi was grateful when able to take a year's leave after an award of a Rockefeller Foundation Professorship to undertake research at Northwestern University in Chicago.

Horace Magoun was born in Philadelphia and grew up in Newport, Rhode Island, where his father was an Episcopal clergyman. Graduating from Rhode Island State College in 1929 with a bachelor's degree in biology, Magoun attended Syracuse University in New York State where he taught zoology and completed his master's degree in 1931. He then began his PhD at the newly created Institute of Neurology at Northwestern University in Chicago under the guidance of leading neurophysiologist Stephen Walter Ranson. These were exciting times to be a doctorate student in Ranson's laboratory for he was just beginning to explore the hypothalamus with stereotaxic surgery – a procedure enabling precise electrode placements, including those that could produce electrolytic lesions, to target any structure in the brain.[2] This instrument was to have a big impact on the development of biological psychology, and Magoun would use it in his doctorate research, submitted in 1934, on the central control of the cat's pupils reflex to light. After his doctorate, Magoun continued working with Ranson, which became a highly productive collaboration resulting in the publication of over thirty papers, and culminating in an extensive 107-page review of the hypothalamus in 1939. Sadly, this relationship was ended with Ranson's sudden death in 1942. It also marked a change in Magoun's research interests, with him now showing a greater interest in anatomy, including tracing the ascending extrapyramidal tracts. Magoun also investigated the disease of polio, which he showed was linked to the damage to the reticular core of the brain stem. His research focus changed yet again in 1947, when Magoun offered space in his laboratory to Donald Lindsley who bought with him an expertise in EEG and in 1948 when Giuseppe Moruzzi arrived as a Rockefeller fellow.

On his arrival at Northwestern, Moruzzi accepted Magoun's invitation to take part in studies on the cerebellum. Their original intent was to electrically stimulate the superior cerebellar peduncle – a major output pathway of the cerebellum, which passes through the lower midbrain to the ipsilateral ventrolateral nuclei of the thalamus. From here, the cerebellar input is able to influence the planning and execution of voluntary movement through its connection with the motor cortex in the frontal lobe's precentral gyrus. Moruzzi and Magoun's intention was to record from the surface of the motor cortex with EEG electrodes after stimulating the peduncle. This experimental procedure, however, produced some unexpected results. More specifically, when Moruzzi and Magoun stimulated the cerebellar peduncle in anaesthetised cats, they found that the EEG responses obtained from the motor cortex changed from a regular slow pattern (characteristic of a sleeping or anaesthetised animal) to one where the EEG was desynchronised and composed of low-voltage fast activity – a response very reminiscent of an arousal reaction in an alert animal. This effect was so unexpected that Moruzzi and Magoun, at first, thought their measurements were faulty – but the change in EEG response turned out to be a genuine consequence of their stimulation.

As Moruzzi and Magoun examined their EEG effect further, they realised that the desynchronised activity (which was obtained with relatively low levels of stimulation) occurred in many cortical regions and was not limited to the precentral gyrus. In other words, their stimulation produced a highly aroused cerebral cortex. And more impressive still, the brain-wave activation outlasted the stimulation by some minutes. At this point, Moruzzi and Magoun began to suspect that their stimulation was spreading beyond the cerebellar peduncles to the surrounding brain stem. Indeed, this would make sense if the fibres of the superior cerebellar peduncle were making connections with the cells of the mesencephalon and pontine reticular formation as they passed to the thalamus. Thus, by stimulating the cerebellar peduncle, it was possible that Moruzzi and Magoun were inadvertently activating the brain's reticular formation. In turn, it was conceivable that this was exciting the cerebral cortex through another thalamic route.

The reticular formation is a complex meshwork of cells and fibres (*reticulum* means net) that run through the brain stem's central core. It had first received serious attention in 1909 when Cajal showed that this part of the brain contained long multiple branching neurons that both ascended and descended through the middle of the brain stem. By the early 1940s, the reticular formation was also recognised as the oldest part of the brain and essential for vital reflexes such as breathing. However, there was little evidence to show at this time that the reticular formation had functional connections with the cerebral cortex. Nonetheless, when Moruzzi and Magoun moved their electrode placements into the reticular formation, they found that stimulation closely mimicked the EEG response obtained from the cerebral peduncles. Furthermore, this effect was evoked from all regions of the reticular formation – extending from the portion of the brain stem lying above the spinal cord (the bulbar region) to the lower parts of the diencephalon where the hypothalamus and thalamus were located. Their initial stimulation of the cerebral peduncles had clearly been a fortuitous mistake.

One important implication of Moruzzi and Magoun's findings was that they directly contradicted Bremer's explanation of sleep. Bremer's theory held that wakefulness depended on peripheral sensory input (visual, auditory, etc.) reaching the cerebral cortex, whereas sleep took place when this sensory information was "turned off" (a process technically referred to as "de-afferentation"). Indeed, Bremer had provided evidence for this theory when he made transactions of the upper brain stem, in an operation known as the *cerveau isolé*. This eliminated nearly all of the sensory input reaching the forebrain, and caused a continuous state of sleep in his laboratory animals. Bremer's theory was also consistent with idea that sleep was a passive state that occurred when there was a reduction in peripheral stimulation as a result of fatigue. Moruzzi and Magoun realised, however, that their findings pointed to an alternative explanation – namely, the reticular formation acted as the brain's arousal centre and it actively implemented the stages of waking and sleep independent of any sensory stimulation. If this was the case, then it was not the sensory pathways governing the control of sleep and waking as Bremer maintained, but the neural connections between the reticular formation and cerebral cortex.

Moruzzi and Magoun examined this possibility further by making lesions to regions of the upper brain stem and then establishing if this damage abolished the effects of reticular formation stimulation on cortical EEG. One region was found to be particularly important in blocking the EEG response. This was an area of the reticular formation that relayed information to the thalamus and basal forebrain called the "mesencephalic tegmentum". Interestingly, the mesencephalic tegmentum had also been damaged in the cerveau isolé lesion made by Bremer – and could provide an explanation for his permanently sleeping animals. From their findings, Moruzzi and Magoun concluded that the brain stem controlled the activity of the cerebral cortex though a system of ascending reticular relays, some involving the thalamus, and whose stimulation could generate high-voltage fast activity slow waves characteristic of sleep and desynchronised fast activity characteristic of arousal and waking. The result of the collaboration between Moruzzi and Magoun was published in the 1949 inaugural volume of the scientific journal *Electroencephalography and Clinical Neurology*.

The passive theory of sleep was not entirely ruled out by Moruzzi and Magoun's work, however, since it was conceivable that their electrical stimulation of the reticular formation indirectly affected the sensory input to the cortex in some way. This problem was addressed by Donald Lindsley, along with Magoun, who made lesions to the reticular formation but which left the sensory pathways from the periphery intact. In this case, Lindsley demonstrated that his animals showed continuous sleeping reminiscent of those found by Bremer with his cerveau isolé preparation. But just as important, when he next made lesions to the sensory pathways, they did not produce somnolence or EEG slow waves predicted by Bremer's theory. Thus, Lindsley et al. (1950) concluded that the loss of arousal and induction of slow wave sleep coma were caused by destruction of the mesencephalic reticular core. Subsequent studies using unanaesthetised behaving animals with electrodes implanted for long-term stimulation and recording also confirmed that the reticular formation exerted rhythmic control over the cortex which was responsible for producing the states of waking and sleep.

The idea of an *ascending reticular formation* where a subset of the reticular formation neurons sends fibres to the intralaminar thalamic nuclei, which in turn project widely throughout the cerebral cortex, is now an important concept in neuroscience. It is also clear this system plays a major role in the regulation of wakefulness and sleep states, although it has also been implicated in many other functions including motor function, learning and memory, emotion and motivation. It is also anatomically complex with ascending fibres utilising a variety of neurotransmitters (noradrenaline, serotonin, acetylcholine, etc.) arising from numerous brain stem nuclei including the locus coeruleus, raphe and pedunculopontine and parabrachial nuclei. There are also contributions to this formation from other thalamic nuclei, hypothalamus and basal forebrain. This large arrangement of neurons is particularly crucial for our very existence as damage leads to a loss of consciousness and coma, even in the absence of any damage to the cerebral hemispheres. Thus, the cerebral cortex requires input from the brain stem to maintain activity and awareness. Without the ascending reticular formation, we would be psychologically dead.

In 1949, Moruzzi returned to Italy to take charge of the institute of Human Physiology in Pisa. Although the laboratories were almost empty, with most of their instruments lost or damaged from the war, Moruzzi transformed the institute into one of the major European neuroscience centres, especially for sleep research, after receiving a Rockefeller grant. He was widely regarded as one of the greatest Italian scientists of the twentieth century, who insisted on working into old age despite the diagnosis of Parkinson's disease in 1975. Magoun remained at Northwestern University until 1950, when he then moved to the University of California at Los Angeles (UCLA) where he founded the Department of Anatomy in the new School of Medicine. This would lead to a fulfilment of an ambition to create a brain research institute in 1959, which occupied a ten-storey building – the first of its kind in the United States. In 1973, Magoun left UCLA to become director of the National Research Council, although he returned to UCLA as emeritus professor in 1974, where he was to remain until his death at the age of 83.

Notes

1 Sadly, Camis who was a Jew, was forced to leave his post in 1938. He would join the Dominican Order as Father Alberto, and in 1939, he would leave for the Philippines to devote himself to medicine and teaching.

2 The stereotaxic frame was first used by Sir Victor Horsley and Robert Clarke to create minute electrolytic lesions in the CNS of animals. It was patented in 1914 and later became adapted for use in human operations.

Moruzzi and Magoun's major writings

Lindsley, D.B., Bowden, J. and Magoun, H.W. (1949) Effect upon the EEG of acute injury to the brain stem activating system. *Electroencephalography and Clinical Neurophysiology*, 1, 475–486.

Lindsley, D.B., Schreiner, L.H., Knowles, W.B. and Magoun, H.W. (1950) Behavior and EEG changes following chronic brain stem lesions in the cat. *Electroencephalography and Clinical Neurophysiology*, 2, 483–498.

Magoun, H.W. (1958) *The Waking Brain*. Springfield, IL: Charles C. Thomas.

Magoun, W.H. (2004) Horace Winchell Magoun 1907–1991. *National Academy of Sciences: Biographical Memoirs*, 84, 249–269.

Moruzzi, G. and Magoun, H.W. (1949) Brain stem reticular formation and activation of the EEG. *Electroencephalography and Clinical Neurophysiology*, 1, 455–473.

Moruzzi, G. (1954) The physiological properties of the brain stem reticular system. In J.F. Delfresnaye., E.D. Adrian., F. Bremer and H.H. Jasper (eds), *Brain Mechanisms and Consciousness*. Oxford: Blackwell.

Further reading

Bremer, F. (1975) The isolated brain and its aftermath. In: F.G. Worden, J.P. Swazey and G. Adelman (eds), *The Neurosciences: Paths of Discovery*. Cambridge, MA: MIT Press.

Hobson, J.A. and Brazier, M.A.B. (eds) (1980) *The Reticular Formation Revisited*. New York: Raven Press.

Marshall, L.H. and Magoun, H.W. (1998) *Discoveries in the Human Brain*. Totowa, NJ: Humana Press.

Sandrone, S. (2013) Giuseppe Moruzzi (1910–1986). *Journal of Neurology*. DOI:10.1007/s00415-013-7036-6.

Shampo, M.A., Kyle, R.A. and Steensma, D.P. (2013) Horace Magoun – Neuroanatomist and neurophysiologist. *Mayo Clinic Proceedings*, 88, e33.

Siegel, J. (2002) *The Neural Control of Sleep and Waking*. New York: Springer-Verlag.

Chapter 11

Donald Olding Hebb (1904–1985)

Hebb's great legacy is his book, The Organisation of Behavior, first published in 1949, which would introduce the concepts of synaptic plasticity, cell assemblies and reverberatory activity to account for the neural events underlying thought and behaviour – ideas that still have a profound impact today on neuroscience and psychology.

Donald Hebb was born in the small fishing town of Chester, Nova Scotia, to parents who were both physicians. As a child, Hebb was a precocious reader and dreamed of becoming a novelist – a pursuit he intended to follow in 1921 when he began studying English and philosophy at Dalhousie University in Halifax. Graduating with a BA in 1925, Hebb then took a teaching course and was employed for a short period as principal at his old school in Chester. Afterwards, he became a farm labourer where he caught tuberculosis. In a state of frail health, he enrolled as a part-time master's student in psychology at Montreal's McGill University in 1928. By now, despite publishing several poems, Hebb's novel writing had not progressed and, having to support himself with teaching, he became head of an elementary school in a working-class district of Montreal (Verdun). But Hebb's teaching career was not to last as he became bedridden for a year (1930–1931), with a hip infection leaving him with a permanent limp. This forced period of recuperation, however, was to have a formative impact on Hebb who managed to read Sherrington's *The Integrative Action of the Nervous System* and Pavlov's *Conditioned Reflexes* (both published in English in 1927). Both the books proved inspirational and led Hebb to write a theoretical master's thesis from his bed on the learning of spinal reflexes in the foetus in 1932 – one of the examiners being Boris Babkin who had collaborated with Pavlov in St. Petersburg.

In 1934, after losing his wife of just 18 months in a car accident, Hebb moved to Chicago to undertake his PhD with Karl Lashley (on the effects of early visual deprivation on size and brightness perception in rats). A year later, Hebb followed his supervisor to Harvard where he completed his PhD in 1936. Although his research was going well, Hebb received an unexpected opportunity to return to Montreal on a Rockefeller Fellowship in 1937 to work with Wilder Penfield who wanted someone to study the psychological effects of frontal and temporal lobe ablation. Although the post was for only 2 years, it was too good to turn down, and Hebb would discover that the removal of large amounts of brain tissue appeared to have little effect on intelligence – a finding leading him to conclude that the results of learning were most likely widely distributed throughout the higher areas of the brain. Hebb followed this experience by taking up a lectureship at Queen's University in Kingston, Ontario. He was to stay in Canada until 1942 when he joined up with Lashley again, now at the Yerkes Laboratory in Florida. At the end of this appointment in 1947, Hebb was offered a professorship at McGill, and within a year, he had become the acting head of its psychology department. He would remain at McGill until his retirement in 1972.

During his time in Florida, Hebb had begun writing the first chapters of his great book *The Organization of Behavior: A Neuropsychological Theory*, which was published in 1949. By now, like Lashley, Hebb had come to the conclusion that stimulus-reflex theories that dominated psychology at the time were inadequate to explain the complexities of behaviour. Thus, he wanted to find a new way of explaining mental function and one consistent with what was known about the brain and neural biology. In this quest, Hebb began to consider many aspects of behaviour that were held to be off limits by the behaviourists, including attention, perception and thinking. This was not an aim fitting well within the zeitgeist of the times that was dominated by B.F. Skinner who denounced anyone who sought to explain behaviour in terms of mental processes or vague physiological processes as "physiologizing". Instead, Skinner held that all behaviour could be explained by the simple laws of conditioning. Hebb did not agree and undaunted by the prospect of criticism he knew would be targeted at him from the behaviourists, set out to show how neural mechanisms, at least in theory, could give rise to a variety of psychological processes that formed the basic constituents of intelligence itself.

One of Hebb's most important innovations in *The Organization of Behavior* was to replace the simple reflex with a neural circuit that contained a large assembly of cells – an idea that had first been suggested by the Spanish neuroscientist Rafael Lorente de No in the 1930s. Thus, instead of viewing the reflex as a one-way chain of neural events that automatically led from the environmental stimulus to the behavioural response, Hebb placed a "loop" in the middle of the neural chain, which increased the number of extra neurons. The reflex arc was now circular (i.e. a cell assembly) and not linear. This concept had a number of advantages over the simple reflex – especially when trying to understand the role of the cerebral cortex where conditioning was believed to take place. For one thing, if set into motion, a cell assembly could produce "reverberations" within the circuit, thereby providing a means of maintaining activity in the cerebral cortex long after the stimulus had terminated. It was a radical but simple change that now enabled the neural activity of the brain to be separated from its sensory input. Of course, this situation did not occur with a simple reflex that always caused an immediate response. And this reverberatory activity also provided a highly plausible explanation for learning since the neural network was dynamic and capable of being strengthened with repeated sensory experience. So, for example, in the case of a baby learning to recognise its milk bottle, the initial sensory experience was hypothesised to establish a new neural circuit, with its repeated stimulation resulting in learning.

But Hebb also realised that the cell assembly could serve other important mental functions. Most importantly, Hebb saw that if the assembly remained active after the sensory stimulation has ceased, then it could also provide the "structural engram" for imagery or mental representations (i.e. ideas and concepts) in the brain. And extending this idea further, Hebb proposed that when sets of cell assemblies became activated in sequence, in what he called a "phase sequence", then this could act as the neural basis for the "temporally organised processes of thinking", otherwise known as the stream of thought. The implications were manifold: activity within the cerebral cortex could now be construed as taking place through large widely distributed networks of neural connections. In effect, Hebb had shown how cell assemblies might give rise to human cognition in a manner consistent with the tenants of neural biology.

Another remarkable innovation of *The Organization of Behavior* was that it pinpointed the actual site in the cell assemblies where learning took place. For Hebb, this was at the junction between neurons – namely, the *synapse*. It is easy to forget just how radical this theory was when Hebb was writing his book in the late 1940s. At the time, synapses were controversial, not least because nobody knew for sure if they existed in the CNS or whether the information they allegedly transmitted was modifiable. In fact, most investigators, Hebb included, would have assumed that the transmission of information across synapses was more likely to be electrical in nature. Hence, it was easy to imagine that bio-electrical fields, and not neurotransmitters, were responsible for synaptic modification – if indeed this was the site of change in the nervous system. In fact, Hebb's book (1949) predates Hodgkin and Huxley's explanation of the action potential in 1952 and Fatt and Katz's quantal theory of neurotransmitter release soon after. And the final proof of synapses would have to wait unto 1954 when the first electron microscope pictures of them appeared, along with the admission by John Eccles that chemical substances provided the only plausible way of accounting for inhibition in neural pathways. Even then, it would have taken a leap of faith to suppose that chemical modification at synapses were involved in learning and memory.

But even more incredible, equipped with only the limited knowledge of synapses at his disposal, Hebb came up with a mechanism to explain how synaptic changes in the cell assemblies could account for learning. To do this, Hebb needed a way to sustain persistent reverberatory activity in cortical circuits, and he derived a rule that he expressed in the following way: *If an axon of cell A is near enough to excite a cell B and repeatedly or persistently takes part in firing it, some growth or metabolic change takes place in one or both cells, so that the efficiency of neural transmission between the two is increased.* This is now known as Hebb's rule and it states that continuous stimulation of a neuron in a circuit will lead to a structural change at its synapse. In turn, this will then increase the likelihood of reverberatory activity being maintained. And if reverberatory activity continued for long enough, some form of permanent memory was likely to result. This was not only an important theoretical development but one that explained how memory could be stored in a network diffusely distributed throughout the brain. If the synaptic connections were strengthened in a cell assembly, then this change could act as memory – which could be elicited again by further stimulation of the circuit.

Although Hebb's ideas concerning cell assemblies and synaptic plasticity were theoretical, they have since become fundamental concepts in neuroscience and psychology. In particular, Hebb's statement of his rule, which has become one of the most quoted sentences in neuroscientific literature, has been adopted as a basic principle in neural networks research and received support from neurophysiological studies. Some of the strongest evidence for Hebb's postulate came in 1973, when Timothy Bliss and Terje Lømo working in London discovered a phenomenon, which later became known as long-term potentiation (LTP). Working with the perforant pathway that projects into the dentate gyrus of the hippocampus, Bliss and Lømo found that its stimulation caused long-lasting voltage changes in its recipient hippocampal neurons. In some cases, these voltage alterations were still present some weeks after the stimulation – proving that the synapses in the pathway had been strengthened by the initial burst of electrical impulses. Since this discovery, LTP has been found in many other areas of the brain, including cerebellum, amygdala and cerebral cortex, and shown to be an important neural substrate underlying learning and memory.

In the introduction of *The Organisation of Behavior*, Hebb writes that "the problem of understanding behavior is the problem of understanding the total action of the nervous system, and vice versa". Consequently, there is much in the book on other aspects of psychology. Indeed, one of Hebb's central concerns was the nature of intelligence and the extent to which it was innate or determinedby experience. Hebb had first examined this issue in his PhD when testing the visual capabilities of animals reared in the dark. This research showed that dark confinement impaired the ability of rats to learn a visual discrimination task – although the animals could overcome their deficit if given an extra period of perceptual learning. This led Hebb to conclude that intelligence consisted of two factors: a fixed or innate component and a variable component that was influenced by the environment. The beneficial effects of early environmental stimulation on later learning was also demonstrated by Hebb when he compared the performance of rats on maze learning who had been reared either as laboratory animals in small cages or as pets at home in a much more enriched setting. The results showed that the richer experience of the pet group made them perform better in maze tasks at maturity – a finding that Hebb believed also generalised to human beings. These findings were to have a significant impact on subsequent research, with other investigators showing that enriched environments led to bigger brains with more synapses.

Hebb, even in his own lifetime, was an extraordinarily influential figure whose influence was widely recognised. His ideas have not only stood the test of time but become engrained in just about every area of psychology and neuroscience. *The Organisation of Behavior* was a truly groundbreaking work that not only played a pivotal role in the demise of behaviourism but also was a central keystone in the cognitive revolution that followed. Interestingly, when writing its early chapters, Hebb offered co-authorship of the book to Karl Lashley who turned down the request. Although Lashley later wrote to Hebb congratulating him on the book, it appears in private he was far less complimentary, and according to one source described it as a garbled version of his own theories (see Glickman 1996). This is certainly not the way history has judged the matter, with one commentator comparing Hebb's book as revolutionary as Darwin's *On the Origin of the Species* (Adams 1980). Indeed, it is hard to imagine a more original or influential book in psychology during the twentieth century. After its publication, Hebb would write on a wide range of matters, including two more books: *A Textbook of Psychology* and *Essay on Mind*. He also served as president of the Canadian and American Psychological Associations and be nominated for the Nobel Prize in physiology and medicine in 1965. Hebb also transformed the psychology department at McGill into one of the most foremost centres for physiological psychology in the world, which would attract a number of high-profile students. After his retirement in 1972, Hebb remained at McGill as emeritus professor and then returned to Dalhousie in the same capacity in 1980. A man who loved sailing in his boat when not working, Hebb died in 1985 at the age of 81 while in hospital for what was expected to be minor corrective surgery on an arthritic hip.

Hebb's major writings

Hebb, D.O. (1947) The effects of early experience on problem solving at maturity. *American Psychologist*, 2, 306–307.

Hebb, D.O. (1949) *The Organization of Behavior: A Neuropsychological Theory*. New York: John Wiley. Reprinted by Lawrence Erlbaum Associates, New Jersey, 2002.

Hebb, D.O. (1958) *A Textbook of Psychology*. Philadelphia: Saunders and Co.
Hebb, D.O. (1974) What psychology is about. *American Psychologist*, 29, 71–79.
Hebb, D.O. (1980) D.O. Hebb. In: G. Lindzey (ed.), *A History of Psychology in Autobiography*. San Francisco: Freeman.
Hebb, D.O. (1980) *Essay on Mind*. New Jersey: Erlbaum Press.
Hebb, D.O. (1982) *Conceptual Nervous System: Selected Papers*. Oxford and New York: Pergamon Press. Edited by H. Buchtel.

Further reading

Adams, P. (1980) Hebb and Darwin. *Journal of Theoretical Biology*, 195, 419–438.
Brown, R.E. and Milner, P.M. (2003) The legacy of Donald O. Hebb: More than the Hebb synapse. *Nature Reviews Neuroscience*, 4, 1013–1019.
Brown, R.E. (2007) The life and work of Donald Olding Hebb, Canada's greatest psychologist. *Proceedings of the Nova Scotian Institute of Science*, 44, 1–25.
Cooper, S.J. (2005) Donald O. Hebb's synapse and learning rule: A history and commentary. *Neuroscience and Biobehavioral Reviews*, 28, 851–874.
Fentress, J.C. (1999) The organization of behavior revisited. *Canadian Journal of Experimental Psychology*, 53, 8–19.
Glickman, S.E. (1996) Donald Olding Hebb. Returning the nervous system to psychology. In: A. Gregory, C. Kimble, A. Boneau and M. Wertheimer (eds), *Portraits of Pioneers in Psychology*. Hillsdale, NJ: Erlbaum Press.
Milner, P.M. (1993) The mind of Donald O. Hebb. *Scientific American*, 268, 124–129.

Chapter 12

Nathaniel Kleitman (1895–1999)

Kleitman is widely regarded as the founding father of sleep research, who not only laid the psychobiological foundations for our understanding of sleep and circadian rhythms but whose endeavours also led to the landmark discovery of rapid eye movement sleep in 1953 with Eugene Aserinsky.

Nathaniel Kleitman was born into a Jewish family in Kishinev, Russia, which now forms the Republic of Moldova, sandwiched between Rumania and Ukraine. This was a period when there were strong anti-Semitic feelings throughout Eastern Europe, and in 1903 and 1905, the city of Kishinev became the scene of violent riots, which saw scores of Jews killed and many hundreds of houses and shops destroyed or pillaged. It was against this backdrop of racial intolerance that the family emigrated to Palestine in 1912 when Kleitman was 17 years old. He began his studies at the Syrian Protestant College (now the American University of Beirut) in Lebanon by taking mathematics and sciences. But the outbreak of World War I terminated his education since Kleitman was still a Russian citizen and therefore a potential enemy. In response, Kleitman fled to the Island of Rhodes in 1915, where, according to one obituary, he boarded the only ship in port, which was bound for New York. Three years later, Kleitman became a naturalised citizen of the United States. Despite speaking poor English, he had managed to enrol in the College of the City of New York soon after arriving and graduated with a BSc in 1920. The same year, Kleitman completed his master's degree from Columbia University, which saw him develop his first interest in the effects of sleep deprivation. After a short spell of teaching physiology and pharmacology at the University of Georgia, Kleitman joined the physiology department at the University of Chicago in 1921, where he wrote a PhD entitled "Studies on the physiology of sleep". Kleitman then spent a couple of years working in Europe, including a stay in Paris where he collaborated with Henri Piéron – author of the first encyclopaedic book on sleep *Le Problème Physiologique du Sommeil* in 1913, and perhaps the only other major physiological psychologist studying sleep in those days.[1] On his return home, Kleitman took up an assistant professorship at Chicago in 1925, where he established a sleep research laboratory by sequestering an old two-room chemistry laboratory, which allowed a bed to be set up in one room and an observer stationed in the other. It was the first of its kind in the world. An early sponsor of Kleitman's laboratory was the Wander Company, manufacturer of the malted milk drink Ovaltine that is still retailed today as an aid for sleep. For the record: after a few trials, Kleitman came to the conclusion that three tablespoons of Ovaltine might enhance sleep – although there was no evidence that two tablespoons were soporific.

Kleitman's first paper on sleep appeared in the *American Journal of Physiology* in 1923 (the same year as he had submitted his master's degree), and it illustrates some of the main themes he pursued throughout his career. In this article, a group of male students were deprived of sleep for periods ranging from 40 to 115 hours, during which time they were subjected to a battery of physiological and psychological tests. The results showed that the sleep loss led to a slowing of heart rate and decline in blood pressure, although the rhythm of core body temperature remained stable. The performance on mental testing revealed that the deprivation did not significantly impair the subject's ability to conduct routine calculations, although their attention often strayed from the tasks. Kleitman also took part in his own experiment and stayed awake for 115 hours. His introspective experience of the sleep deprivation experience indicated that a greater likelihood of delusional thoughts was a distinct possibility of prolonged insomnia. Later, Kleitman performed sleep deprivation studies on animals, including young puppies, and showed that it caused a change in immune function with a significant drop in the production of red blood cells.

During the 1930s, Kleitman's interest in sleep expanded into new areas. This can readily be seen in his great book, published in 1939 and entitled *Sleep and Wakefulness as Alternating Phases in the Cycle of Existence*. This monograph was a comprehensive survey of sleep research up to that point, along with related subjects such as hypnosis and hibernation. It also included Kleitman's translation of papers that had been originally published in French, German, Italian and Russian – studies largely unknown to the English-speaking world. A revised and larger edition of the book also appeared in 1963, with 552 pages, 36 chapters and a list of 4,337 references in the bibliography. Apparently, when Kleitman first wrote *Sleep and Wakefulness*, in an era when references could only be adequately stored on cards, the bibliography took him longer to compile than the writing of the book itself. Over 50 years later, the book is still widely referred to as the "the bible of sleep researchers".

One of the great unsolved mysteries of sleep research at the time of *Sleep and Wakefulness* was whether the 24-hour circadian human sleep-wake rhythm is learned through habit or whether people had an internal hard-wired body clock. To investigate this, in 1938, Kleitman undertook one of his most famous experiments, when he and an American colleague Bruce Richardson, spent 32 days alone, some 400 metres underground, in Mammoth cave in Kentucky. The cave was chosen because it had no natural light, or sound from outside, and was a constant and chilly 54° F. Lanterns were used to regulate light and dark, and the cave also contained a table, washstand and two beds on extended legs to counter the damp floor and rats. A concession to modern living was a cook who made regular food deliveries. The main purpose of the ordeal was to discover if the two men could adapt to a 28-hour day – and to this end, for each "day" they slept 9 hours, worked 10 hours and then rested for another 9 hours. During this time, Kleitman and Richardson measured the daily rhythms of their body temperature.

The results showed that Kleitman, who was then 43 years old, was unable to fully adjust to the 28-hour day – his sleep being intermittent and shallow, especially when it coincided with noon in the real Kentucky time outside. His body temperature also stubbornly followed the rhythms of the world that existed outside the cave, which led it to run in and out of synchronisation with his sleep patterns. Richardson, on the other hand, began to adapt to his temperature cycle and sleep pattern after 1 week. Although the study was not conclusive, the Mammoth Cave experiment lent support to the idea that human bodies contain a "clock" that maintains a roughly 24-hour temperature cycle, even in the absence of external cues. Later in the 1970s, experimenters would show that animals, including humans, do indeed have an internal clock, called the "suprachiasmatic nucleus", located in the hypothalamus, which controls the release of the sleep-promoting hormone melatonin and is recalibrated on a daily basis by the occurrence of light.

Kleitman continued this line of enquiry in 1948, when he spent 2 weeks aboard a submarine, the USS *Dogfish*, examining the relationship between mental alertness and temperature change – a situation that revealed that, when core body temperature increased, so did performance. Later, in 1951, he travelled to Norway to conduct research on the stability of the circadian rhythms of sleep in the continuous summer light of the arctic summer. Assisted by his daughter Hortense, Kleitman examined the duration of sleep taken by citizens of Tromsö, which lay north of the Arctic Circle. He found that adults took on average 7.2 hours of sleep per 24-hour period, with a range of 6.1 to 9.4 hours. This was about the same as a similar sample of adults living in Chicago during the same summer. These studies would also lead Kleitman to recognise that humans have a regular pattern of rest and activity in both sleep and wakefulness that lasts about 90 minutes – something he called the "basic rest-activity cycle".

Despite these many achievements, Kleitman is best known for his co-discovery of rapid eye movement (REM) sleep, which he made in 1953 with his PhD student Eugene Aserinsky. Aserinsky is an unlikely candidate to be included in the list of great neuroscience discovers. He had begun his college education majoring in social studies but switched to Spanish before deciding to follow medicine. After 3 years without achieving any qualification, Aserinsky then entered dental school at the University of Maryland in Baltimore – but never took his final exams. Drafted into the army in 1943, the same year as his son Armond was born, Aserinsky was sent to England where he served as a high explosive handler until 1945. On his return to Baltimore, he was employed as a social worker and supervisor in Maryland's department of employment. Deciding to complete his education by making use of the GI Bill (this statute of 1944 provided a range of benefits for returning World War II veterans, which included the payment of tuition fees for further education), Aserinsky enrolled as an undergraduate student in physiology at the University of Chicago – although did not complete his degree. Despite this setback, in 1949, Aserinsky approached Kleitman to ask if he could work in his laboratory. Even then, Aserinsky would write in a later memoir that sleep was the least desirable of the scientific areas he wanted to pursue. But Kleitman took him on.

At the time of Aserinsky's arrival, Kleitman had become interested in the possibility that blinking was an indicator of tiredness and sleep onset. Thus, he suggested to Aserinsky that this subject might be an interesting one to examine in infants. Although this would lead to months of tedious observation, it did show something more interesting: sleeping babies had roughly 20-minute periods each hour without eyelid movements. It was a simple observation but sufficient for Kleitman to suggest that Aserinsky enrol for a PhD despite having no suitable academic qualifications. Aserinsky recounts that the doctoral committee at Chicago was nonplussed at Kleitman's act of faith, as it was essentially allowing him to jump from high school to a PhD degree! Nonetheless, Aserinsky became a PhD student and began to examine whether adults had the same 20-minute sleep episodes as those in infants. He also decided to measure eye lid movements using the electrooculogram (a machine that records the difference in electrical charge between the front and back of the eyes) and brain waves from the scalp with the electroencephalogram (EEG).

After coming to grips with the technical hitches of his equipment, Aserinsky found to his surprise that when his subjects fell asleep, they soon entered a state where they exhibited rapid eye movements (REM). Curiously, this state also coincided with fast wave brain electrical activity that was more typical of a subject who was awake and alert. At first, Aserinsky and Kleitman were inclined to believe that these two measures were associated with some form of instrumentation defect. Nonetheless, when they woke their subjects up during a period of REM sleep, they nearly always reported vivid dreaming – something that rarely occurred in the non-REM periods of sleep. Only then, did it become apparent that Aserinsky and Kleitman had discovered two completely but alternating different sleep states, which are now commonly referred to as slow wave sleep (SWS) and REM sleep. These findings, first published in *Science* in 1953, were revolutionary. Some 4 years later, another student who had joined Kleitman's laboratory, William Dement, would show that the alternation of SWS and REM sleep took place in regular bouts that lasted 90 minutes – with an average night's sleep having four or five of these cycles.

Some people still express surprise at Aserinsky and Kleitman's discovery – not least because it seems obvious now that REMs are associated with dreaming. Nor was this type of research a young discipline at the time since the first sleep and waking EEG rhythms had been reported by Hans Berger in 1925. So why had nobody recognised REM sleep earlier? The explanation, it appears, is that researchers had not recorded for long enough, or instead had noticed patterns of unusual EEG activity in their subjects during sleep, but assumed that they represented brief periods of awakening. Thus, Aserinsky was the first person who was attentive enough to notice they occurred in subjects who were actually asleep. As the distinguished sleep researcher Jerome Siegel has commentated, there is no substitute for careful observation in experimental research!

Kleitman remained at Chicago until his retirement in 1960 when he was made professor emeritus. He then moved to California where one of his daughters lived, although he continued writing and research and made frequent trips back to Chicago, publishing his last paper at the age of 87. Even at this grand old age, Kleitman was happy to participate in research, becoming a subject in a longitudinal study on sleep and ageing from his eighty-sixth to ninety-forth year. His book *Sleep and Wakefulness* continues to be the standard reference on the subject. It has not only laid the foundation for establishing the study of sleep as a laboratory-based discipline but also helped raise awareness of circadian rhythmicity as an important public health issue. Consequently, today, we know much more about the consequences of shift-work and jet lag, along with various sleep disorders such as insomnia, narcolepsy and sleep paralysis, because of Kleitman's pioneering work. He died in 1999 at his home in Los Angeles, California, aged 104 years. Kleitman had survived Aserinsky by 1 year, who was killed in 1998 at the age of 77, when his car hit a tree close to San Diego. Although an autopsy proved inconclusive, it is possible that the accident had resulted from him having fallen asleep at the wheel.

Note

1 Piéron was also noted for his hypnotoxin theory of sleep – the idea that the brain produces its own endogenous sleep substances.

Kleitman's major writings

Aserinsky, E. and Kleitman, N. (1953) Regularly occurring periods of eye motility and concomitant phenomena, during sleep. *Science*, 118, 273–274.

Dement, W. and Kleitman, N. (1957) Cyclic variations in EEG during sleep and their relation to eye movements, body motility, and dreaming. *Electroencephalography and Clinical Neurophysiology*, 9, 673–690.

Kleitman, N. (1923) The effects of prolonged sleeplessness in man. *American Journal of Physiology*, 66, 67–92.

Kleitman, N. and Kleitman, H. (1953) The sleep-wakefulness pattern in the Arctic. *The Scientific Monthly*, 76, 349–356.

Kleitman, N. (1963) *Sleep and Wakefulness*. Chicago: University of Chicago Press.

Kleitman, N. (1982) Basic rest-activity cycle – 22 years later. *Sleep Research and Sleep Medicine*, 5, 311–317.

Further reading

Aserinsky, E. (1996) Memories of famous neuropsychologists: The discovery of REM sleep. *Journal of the History of the Neurosciences*, 5, 213–227.

Dement, W.C. (1974) *Some Must Watch While Some Must Sleep*. New York: Freeman.

Dement, W.C. (2001) Remembering Nathaniel Dement. *Archives Italiennes de Biologie*, 139, 11–17.

Foster, R. and Kreitzman, L. (2005) *Rhythms of Life*. London: Profile Books.

Kroker, K. (2007) *The Sleep of Others and the Transformations of Sleep Research*. Toronto: University of Toronto Press.

Lamberg, L. (2004) The student, the professor and the birth of modern sleep research. *Medicine on the Midway*, Spring, 16–25.

Lavie, P. (1996) *The Enchanted World of Sleep*. London: Yale University Press.

Pollak, C.P., Thorpy, M.J. and Yager, J. (2009) *The Encyclopedia of Sleep and Sleep Disorders*. New York: Chelsea House Publishers.

Chapter 13

Alan Lloyd Hodgkin (1914–1998) and Andrew Fielding Huxley (1917–2012)

Although they only worked together for a few years, Hodgkin and Huxley formed one of the most influential partnerships in neuroscience, whose main achievement was a series of five papers, published in 1952, which fully described the biochemical and physiological basis of the action potential.

Alan Hodgkin was born in the small English town of Banbury, Oxfordshire, into a strict Quaker family, shortly before the outbreak of World War I. Although both his grandfather and uncle were historians, as a boy, Hodgkin had an abiding passion for wildlife, especially ornithology, and would win a medal at the age of 16 for an essay on bird behaviour from the Royal Society for the Protection of Birds (RSPB). In 1931, Hodgkin won a scholarship to Trinity College, Cambridge, and the following year, he began his Natural Science Tripos degree. Although intending to make zoology his preferred subject, he also decided to learn some additional physics and mathematics. By his second year, however, Hodgkin's interest had shifted to physiology and he became involved in the running of his own experiments. In fact, his first foray into physiological work examined nerve conduction – research he continued after obtaining his first-class degree in 1935. This would lead to Hodgkin winning a Trinity College research fellowship in 1936 and a Rockefeller Fellowship in New York the following year, where he worked with the Nobel winning neurophysiologist Herbert Gasser. But perhaps even more importantly for Hodgkin's career, he also visited Kenneth Cole and Howard Curtis at the Wood Hole Marine Biological Laboratory in Massachusetts where he was introduced to the giant nerve fibre of the squid (see below). On his return to England, Hodgkin continued his experiments at the Plymouth Marine Biological Laboratory where he had also worked as a schoolboy.

Andrew Huxley was born in Hampstead, London, into a highly illustrious academic family. His father was the writer Leonard Huxley and his grandfather was Thomas Huxley, the distinguished nineteenth-century biologist and vigorous advocate of evolutionary theory, who was widely known as "Darwin's bulldog". In addition, the biologist Sir Julian Huxley and Aldous Huxley, the author of *Brave New World*, were half-brothers from his father's first marriage. Huxley therefore had an exemplary education in more ways than one. After studying at University College School in Hampstead (1925–1930), Huxley went to Westminster School (1930–1935) where he came under the influence of J.F. Rudwick who inspired him to turn from his love of classic literature to science. In 1935, Huxley entered Trinity College, Cambridge, after winning an Entrance Scholarship, where he soon became enticed by physiology. He spent from 1937 to 1938 studying anatomy with the intention of qualifying in medicine and then focused on physiology in the final part of his Natural Sciences Tripos degree (1938–1939). His introduction to physiological research began in 1939 when he joined Alan Hodgkin, who had previously tutored him at Trinity, in Plymouth.

By the late 1930s, much had been learned about the nature of nervous conduction. The all-or-none principle had been established by Douglas Adrian and the nervous system shown to use a form of frequency coding to encrypt information. These were big steps forward. However, a central question remained unsolved: namely, the electro-chemical events that produced the nerve impulse (otherwise known as the *action potential*). This ignorance was hardly surprising. The tiny size of neurons and their axons, which in mammals are no bigger than a few microns (thousandths of a millimetre) in diameter, meant it was not possible to record voltage changes inside the cell. Nor were researchers able to measure the minute electrical currents passing through the membrane during the nerve impulse. Both required probes small enough to be placed inside and outside the cell. Somewhat unexpectedly, the answer to these difficulties would be reported to the scientific world in 1936 by Oxford biologist John Zachary Young, who was an expert on the nervous system of cephalopod molluscs – a group that included the octopus, squid and cuttlefish. Young had been working in Plymouth when he noticed that the mantle cavities of the squid contained large transparent tubular structures. At first, he thought these were blood vessels but came to realise they were neural ganglion. Importantly, the axons making up these structures were about a millimetre in diameter – that is, over a hundred times larger than those found in mammals. This was the breakthrough eletro-physiologists had been hoping for. Electrodes, or fine wires, could be implanted in the axon's interior, and it was also possible to "squeeze" out the cytoplasm to determine its chemical composition. The axons were also relatively simple to dissect out from the squid's mantle and keep alive in a bath of salt water for several hours.

In the summer of 1939, Cole and Curtis working in Massachusetts and Hodgkin and Huxley in Plymouth independently began examining the electrical properties of the giant squid axon. The basic procedure adopted by both two groups involved inserting an extremely fine recording electrode into the axon while placing a second one on the surface, so its tip was just across the membrane. This allowed the voltage differences at the two sites to be measured during an experimentally induced nerve impulse (action potential) passing down the axon. One of the first results obtained by Cole and Curtis was that the axon's membrane reduced its electrical resistance some 40-fold when an action potential reached the recording point – which is another way of saying the membrane became forty times more permeable at that moment to the passage of electrically charged ions (an ion is an atom, such as sodium or potassium, which has lost or gained an electron, which then makes it positively or negatively charged).[1] The "breakdown" of the membrane was also very brief – taking place over a period of just 3 to 4 msecs.

Hodgkin and Huxley also examined the voltage differences inside and outside the giant squid's axon during its resting state (i.e. when no nerve impulses were occurring). This too produced an unexpected finding: the axon's interior charge was about –45 mV compared to the outside. Although it has been known since the nineteenth century that the inside of the resting nerve cell was negative, this disequilibrium was far greater than had been expected. In fact, it was a voltage difference that placed the axon under considerable electrical tension – especially from positive sodium ions in the extracellular fluid, which were electrically attracted to the axon's negatively charged interior (a positive charge is attracted to negative and vice versa).

Hodgkin and Huxley then examined the voltage changes during an action potential. It was widely believed that the voltage inside the cell would become zero as the membrane broke down and became more permeable to ions. But the current flow into the axon was found to be much stronger, with Hodgkin and Huxley recording an overshoot of about +45 mV. Moreover, when the action potential was plotted over time, it could be seen to be an event lasting no more than 2 to 3 msecs, with the voltage inside the axon jumping sharply from about –45 mV at rest to about +40 mV at its peak, before dropping back to its resting value. The overshoot of electrical charge inside the axon also indicated there was some sort of "extra" process at work, other than a passive breakdown of the membrane.

Unfortunately, further work had to be put on hold because of the outbreak of World War II in 1939. Hodgkin turned his attention to the development of radar, while Huxley worked on navel gunnery. Even when the war was finished, the resumption of scientific activities was slow. There were fuel and food shortages in Britain and the Plymouth laboratory had been damaged by German bombs. Consequently, Hodgkin and Huxley did not renew their partnership until 1947, when they began to consider why a marked reversal of membrane potential (negative to positive) occurred during an action potential. One possibility was that axon's membrane potential becomes positive through the sudden inflow of positive sodium (Na+) ions. Indeed, these ions were found in high concentrations in the body's extracellular fluids where they helped maintain the electrolytic balance. Moreover, by the 1940s, researchers knew that the excitation of muscle depended on the inflow of Na+ ions. If this occurred for muscle, why not the nerve axon? In 1947, Hodgkin tested this idea with pharmacologist Bernard Katz. When they removed the Na+ ions from the fluid bathing the axon, they found this abolished the action potential. In contrast, increasing Na+ levels made the axon more excitable. It was strong evidence that the flow of Na+ ions into the axon was the mechanism of depolarisation that accompanied the action potential.

During their investigations, Hodgkin and Huxley also discovered another fact that proved crucial for understanding the action potential. They showed that the nerve impulse was only initiated, or set into motion down the axon, if the cells resting potential (which they had shown was always around –45 mV) was increased to about –30 mV. Put another way, if the voltage inside the cell, or more specifically at the axon hillock (where the axon joins the cell body), was increased by about 15 mV, this signal acted as the trigger for the full-blown action potential. The +15 mV "threshold potential" was therefore the critical point at which the membrane would lower its resistance to Na+ ions, thereby setting into motion the train of events reversing the voltage inside the cell from negative to positive.

But the real key to unlocking the nerve impulse was to measure the concentration of all ions, and their inward and outward currents, on a millisecond-by-millisecond basis. Even with the giant squid axon, this was initially impossible because the action potential was too fast and currents too small. However, the problem was solved by the "voltage clamp" technique first used by Kenneth Cole in 1949. This procedure used a system whereby two electrodes were implanted inside the axon – one measuring the voltage inside the cell and the other producing a current that was able to alter this to a level determined by the experimenter. Thus, the technique enabled the researcher to set the voltage potential inside the cell and then to maintain it (i.e. "clamp it") without necessarily triggering the action potential. Although this is an artificial situation, it allows the current flow across the membrane to be accurately recorded at the clamped point. Alternatively, the clamping can be used to instantly halt the action potential at any moment, thereby allowing the observer to measure the flow of current and estimate ion concentrations necessary to produce the voltage changes being observed.

From 1948 to 1951, Hodgkin and Huxley used the voltage clamp technique in conjunction with other procedures, such as varying the ion concentrations surrounding the axon or by applying drugs known to interfere with the movements of specific types of ion across the membrane, to elucidate the action potential. Perhaps their most important finding was that the action potential does not travel down the core of the axon, but rather along the outer membrane, as a cascading wave of inward flowing of NA+ ions to create a positive current, followed by the outward flow of potassium (K+) ions, which brings the resting potential back to negative. In addition, Hodgkin and Huxley also began to realise that the membrane did not passively "breakdown" to allow the passage of these ions. Rather, there had to be special channels that "actively" transported the NA+ and K+ ions across the membrane to allow for these two different but highly synchronised currents to occur.

These were groundbreaking experiments, although Hodgkin and Huxley did not stop there, for they used their data to create a series of mathematical equations to explain and describe the action potential. How these equations were calculated has become part of physiological folklore. The initial analysis took place in the cold winter of 1947 when Huxley, unable to use the Cambridge University's only computer, was forced into using a manual calculator with mitten-covered hands. The task took him about 6 months to complete (Huxley later recalled that it took him around 8 hours of solid computation just to plot a 5 msec description of a given variable) and it resulted in what has been called the most important theoretical achievement in modern neurobiology. Hodgkin and Huxley's equations were published in a series of five landmark papers in the *Journal of Physiology* in 1952, and they outlined all the key biophysical properties of the action potential, including its form, duration, amplitude, velocity and ionic movements. The explanation of the action potential derived from these equations was so accurate that it has never been seriously challenged.

For their brilliant work, Hodgkin and Huxley were awarded with numerous awards including the Nobel Prize for physiology or medicine in 1963, along with knighthoods (Hodgkin in 1972 and Huxley in 1974). Somewhat surprisingly, despite both remaining at Cambridge in the wake of their seminal research, they would not collaborate again. Hodgkin continued his investigations into the giant squid axon, before moving on in the 1960s to examine ionic currents in the photoreceptors of the eye. He would also serve as president of the Royal Society from 1970 to 1975. Huxley turned his attention to the conductance of electric current in muscle and was appointed professor at University College, London, in 1960, a post he held until 1983. He would retire in 1990 and live in Grantchester, where his hobbies included building microscopes and other scientific instruments. Hodgkin remained at Cambridge, actively involved in research until about 1987 when he started writing his autobiography *Chance and Design: Reminiscences of Science in Peace and War*. Both men would also enjoy long lives: Hodgkin dying at the age of 84 in 1998 and Huxley at the age of 94 in 2012.

Note

1 The flow of ions across the membrane, in effect, creates a simple form of electric current.

Hodgkin and Huxley's major writings

The five classic papers mentioned in the text appeared in the *Journal of Physiology*, Volumes 116 and 117 (1952).

Hodgkin, A.L. (1963) The ionic basis of nervous conduction. In: *Nobel Lectures: Physiology or Medicine 1963–1970*. Amsterdam: Elsevier (1972).

Hodgkin, A.L. (1992) *Chance and Design: Reminiscences of Science in Peace and War*. Cambridge, UK: Cambridge University Press.

Hodgkin, A.L. (1996) Sir Alan L. Hodgkin. In: L.R. Squire (ed.), *The History of Neuroscience in Autobiography, Volume 1*. Washington, DC: Society for Neuroscience, pp. 252–292.

Huxley, A.F. (1963) The quantitative analysis of excitation and conduction in nerve. In: *Nobel Lectures: Physiology or Medicine 1963–1970*. Amsterdam: Elsevier. (1972).

Huxley, A.F. (2002) From overshoot to voltage clamp. *Trends in Neurosciences*, 25, 553–558.

Huxley, A.F. (2004) Andrew F. Huxley. In: L.R. Squire (ed.), *The History of Neuroscience in Autobiography, Volume 4*. Amsterdam: Elsevier, pp. 214–318.

Further reading

Häusser, M. (2000) The Hodgkin-Huxley theory of the action potential. *Nature Neuroscience Supplement 3*, 1165.

Levy, A. (2013) What was Hodgkin and Huxley's achievement? *British Journal for Philosophy of Science*, 65, 469–492.

McComas, A.J. (2011) *Galvani's Spark: The Story of the Nerve Impulse*. Oxford: Oxford University Press.

Meunier, C. and Segev, I. (2002) Playing the devil's advocate: Is the Hodgkin-Huxley model useful? *Trends in Neurosciences*, 25, 558–563.

Piccolino, M. (2002) Fifty years of the Hodgkin-Huxley era. *Trends in Neurosciences*, 25, 552–553.

Schwiening, C.J. (2012) Classical perspectives: A brief historical perspective: Hodgkin and Huxley. *Journal of Physiology*, 590, 2571–2575.

Chapter 14

Paul Donald MacLean (1913–2007)

MacLean is best known for identifying the limbic system (a term he coined) with "visceral" emotional behaviour and for his theory that the mammalian brain has three evolutionary distinct anatomical parts, each with their own behavioural repertoire.

Paul MacLean (pronounced mac-LANE) was born in the small town of Phelps, upstate New York, to a Presbyterian minister father and a church-going mother who sang in the choir. The third of four sons, MacLean's early schooling took place in Batavia, New York, where by the age of 8 he had already cultivated a love of painting. By his teens, MacLean had acquired a taste for English literature and won a scholarship to become a boarder at the Taft School in Watertown, Connecticut, which prepared many of its pupils for Yale. Following one of brother's footsteps, MacLean entered Yale in 1931 where he took a wide range of courses, although he was always intent on majoring in English. MacLean also took a liking to philosophy and decided that once he had obtained his BA he would spend a year at the University of Edinburgh studying under the Plato scholar A.E. Taylor. However, before setting off for Scotland, in 1935, MacLean's mother had a suspected heart attack (it turned out to be a gall stone) requiring the attention of a local physician. MacLean was so impressed by the doctor's expertise and capacity to bring relief that he made up his mind to pursue medicine. Thus, at Edinburgh, MacLean mixed philosophy with physics and chemistry, enabling him to enter Yale medical school in 1936. Graduating with his MD in 1940, MacLean specialised in cardio-vascular disease and spent his internship at John Hopkins working in an emergency ward. A year later in 1941, the United States entered World War II, with MacLean joining Yale's 39th General

Hospital Brigade, serving in New Zealand where he treated soldiers who had been evacuated from the Solomon Islands. Surprised to find himself as the only person with experience in bacteriology, MacLean was able to spend some of his service undertaking basic research, which showed the diphtheria bacillus to be the cause of tropical ulcers – a discovery that paved the way for effective prophylaxis and treatment. But a more formative experience for MacLean occurred when the resident psychiatrist of his hospital disappeared, which resulted in him being assigned to treating psychiatric patients. It was MacLean's first introduction to emotional disorders such as shellshock and psychosis – and it even led to him being attacked by one of the patients who tried to strangle him with his own necktie!

After leaving the army in 1946, MacLean took up a private medical practice in Seattle, combined with a clinical appointment at the University of Washington Medical School. But he quickly became disenchanted with his new routine when he realised that many of his patient's symptoms were psychological in nature and not something that could be properly treated by what then had become the standard remedy – a sedative. Consequently, he sought a return to research, and it led him to Stanley Cobb, an authority on psychosomatic medicine and founder of the psychiatry department at Massachusetts General Hospital in Boston. Here, MacLean began a research project assessing the bioelectrical activity of the anterior medial temporal lobe region that includes the hippocampus. At the time, this region was attracting considerable attention for its involvement in psychomotor epilepsy – first described by the American neurologist Frederic A. Gibbs in the late 1930s. This type of epilepsy is most commonly associated, not with grand mal seizures, but impaired awareness including amnesia (sometimes called an "absence") in which the person exhibits automated lip-smacking and repetitive movements of the hands and fingers. In addition, psychomotor epilepsy is often preceded by an aura that can give rise to a wide range of highly charged emotions including euphoria, sadness, fear or anger. MacLean was particularly interested in establishing what parts of the temporal lobes were triggering these emotions and designed nasopharyngeal electrodes that could be slipped up the sides of the nose, which lay only about 2 cm from the tissue of the anterior medial temporal lobe. These electrodes would show the seizure activity and emotionally charged auras most often arose from areas surrounding the hippocampal formation.

In the late 1940s, the behavioural functions of the hippocampus were still a mystery. In fact, many researchers still referred to the hippocampus as the rhinencephalon (nose-brain) due to its close anatomical connections with the olfactory system. Clearly, there was much more to the hippocampus than this, and MacLean began to look around for new clues to confirm its involvement in other behaviours. This would lead MacLean to come across a relatively obscure 1937 paper by the respected comparative neurologist James Papez, which identified the hippocampus as an important contributor to the processing of emotion. At the time, it was the hypothalamus that most researchers recognised as the most important brain structure involved in the expression of emotional behaviour. But Papez pointed out in his paper that there had to be somewhere in the brain where the psychological experience of emotion took place. The likeliest site for this response, he argued, was the cingulate gyrus – an old part of the cerebral cortex lying under the main cortical mantle. To explain the interaction of the hypothalamus with the cingulate gyrus, however, Papez assigned an important role to the hippocampus. More specifically, he reasoned that the hippocampus must receive information from the cingulate gyrus, which then sends emotional input, via the fornix, to the posterior region of the hypothalamus. In turn, this area sends a pathway to the anterior nucleus of the thalamus, which then projects to the cingulate gyrus, thereby completing the circuit. Viewed in this way, the hippocampus was important for establishing the psychological emotional response – a theory, which from MacLean's perspective, accounted for the emotive nature of the auras in psychomotor epilepsy.

In 1948, MacLean managed to spend a few days learning neuroanatomy in Papez's Cornell University laboratory – a visit he would describe as life-changing. MacLean was also intent on substantiating Papez's theory with new neuroanatomical and behavioural evidence – and in 1949 published a paper purporting to show how his circuit was linked with various forms of psychomatic illness. Although MacLean added some new structures to the Papez circuit, including the septum and amygdala, his most significant innovation was to replace the old term "rhinencephalon" with a more appropriate *visceral brain*. This was an inspired change in terminology because visceral literally meant "relating to deep inward feelings rather than the intellect" (from the Latin *visceralis* meaning entrails or inner organs of the body). And it was a term that now encapsulated the idea the hippocampus and its system of connected structures functioned as the brain's emotional centre.

MacLean also made a number of new important claims that bolstered and expanded this view. In particular, he regarded the visceral brain as an evolutionary old part of the mammalian brain with close ties with the autonomic nervous system, which also acted as the highest coordination centre for controlling the basic drives of the organism such hunger, reproduction and so on. Thus, the emotional visceral brain contrasted with the more recent neocortex that held sway over the body's musculature and intellect. In addition, MacLean argued that while the visceral brain is conscious, it nonetheless "eludes the grasp of the intellect" because of its "animalistic and primitive nature". It was clear, therefore, that MacLean regarded the visceral brain as having its own mind, and one that was largely segregated from higher regions that supported reasoning and logical thought.

In the same year as he published his paper, MacLean moved to the medical school at Yale. Although the use of the word "visceral" had drawn attention to the brain's emotional centres, it attracted criticism from physiologists who used the word in a more limited way. For them, visceral was a term that only applied to the glands and hollow organs of the body such as the blood vessels. To overcome this problem, MacLean began using the expression *limbic system* instead. It was an interesting change. The Latin word limbus means "on the border" and had entered anatomy during the seventeenth century. But more importantly, the term *le grand lobe limbique* had been adopted by the great French doctor Paul Broca in 1877 to describe the cingulate and parahippocampal gyri that encircle the limbus (or medial boundaries) of each hemisphere. Broca had also recognised this lobe was a consistent feature of all mammalian brains, although it had "varying degrees of distinctness, size and completion". Despite this, Broca did not believe it was important for human behaviour, noting that in more intelligent species there seemed to be atrophy of the limbic lobe at the expense of the far greater development of the frontal lobes. Nor could Broca attribute the limbic lobe with any meaningful function, except for olfaction and the "bestial sense". According to MacLean, Broca's limbic lobe had therefore become "the unwanted child of brain anatomy books and lectures" because of its perceived irrelevance to human behaviour.

It is also interesting to note at this point, Papez had made only bare mention of Broca's limbic lobe in his 1937 paper, and MacLean none at all in his 1949 article. Consequently, when MacLean first used the term limbic system in 1954, which he defined as "the great limbic lobe of Broca and its subcortical cell stations", this largely overlooked area of the brain began to take on a new lease of life. It may have been a small change of terminology, but it would have a big impact for MacLean had ushered in a new contemporary concept – the idea of an integrated "system" dedicated to serving emotional functions rather than a set of individual structures. Today, the concept of a limbic system is so well engrained in neuroscientific thought that it is easy to forget that the term has such a recent origin.

In 1957, MacLean was invited to head a new department at the National Institute of Mental Health (NIMH), which had been specifically created to examine the limbic system's role in behaviour. By now, MacLean was also becoming interested in the evolutionary development of the limbic system across different species, and to explore this subject further, in 1959, he established a 30-acre field laboratory in Poolesville, Maryland, Here, MacLean undertook naturalistic observation and experimental work on all three classes of terrestrial vertebrates: reptiles, birds and mammals. This work also led to MacLean's other great idea – the notion of the *triune brain* – which he first outlined in a lecture series in 1969. However, it would receive a more comprehensive treatment in his book *The Triune Brain in Evolution* (1990). Tracing brain evolution from reptiles to mammals, MacLean argued this ancestral journey had caused man to have three types of brain, each with their own set of behavioural capabilities. The oldest is reptilian in nature (the *proto-reptilian formation*) and consists of the brain stem, cerebellum, corpus striatum and some other nuclei. For MacLean, this part of the brain supported life and also gave rise to primitive instincts such as exploration, feeding, aggression and sexuality. Over the eons of evolutionary time, however, this brain region became subsumed under the *paleomammilian formation*, or what can be regarded as the limbic system. Believing this first came into existence before the evolution of primates, it was seen as mediating emotional forms of behaviour including those linked with social attachment including separation distress, social bonding, playfulness and maternal protection. Finally, MacLean called the most recent part of the brain the *neomammalian formation*, which corresponds with the neocortex. This reaches its greatest complexity in humans where it overshadows the rest of the brain and is responsible for our all higher cognitive functions. It is also interesting that MacLean's concept of the triune brain with its reptilian, limbic and neomammalian components has some interesting parallels with Freud who postulated the existence of the id, ego and super ego as important components of the human mind.

There is little doubt that MacLean was instrumental in establishing the limbic neurobiology of emotions as an important subject in neuroscience – and his work continues to exert a significant effect today. By introducing the concept of an old mammalian layer in the human brain called the "limbic system", which mediates our affective behaviour and primary social emotions, MacLean has highlighted the fact that we are the product of a long evolutionary journey, with the vestiges of this process most likely still active in our brains. This viewpoint has many ramifications for explaining human behaviour – not least in the field of psychiatry where an understanding of the limbic system's involvement in emotion is likely to provide important insights into many forms of mental illness. The concept of the limbic system is also one known to every student of neuroscience, although it should be noted that MacLean's theory has attracted some criticism. For example, commentators have pointed out that there is no universally agreed upon definition of what constitutes the limbic system and that it is more likely to exist in the imagination of certain anatomists than on true empirical grounds. In fact, there are even some experts such as Joseph LeDoux who believe the term should be obsolete and abandoned. Nor can the limbic system be regarded simply as the visceral or emotional part of the brain, since there is much evidence to show that it is involved in many other different functions including memory, the mapping of our spatial world, navigation and face recognition (to name a few). Nonetheless, despite his critics, MacLean's theories of the limbic system and triune brain are strangely compelling and still shape our perspective today. The neuroscience of emotion has come a long way since MacLean formulated his limbic system concept in 1954.

MacLean's major writings

MacLean, P.D. (1949) Psychosomatic disease and the "visceral brain": Recent developments bearing on the Papez theory of emotion. *Psychosomatic Medicine*, 11, 338–353.

MacLean, P.D. (1954) The limbic system and its hippocampal formation: Studies in animals and their possible application to man. *Journal of Neurosurgery*, 11, 29–44.
MacLean, P.D. (1985) Evolutionary medicine and the triune brain. *Psychological Medicine*, 15, 219–221.
MacLean, P.D. (1990) *The Triune Brain in Evolution: Role in Paleocerebral Functions*. New York: Plenum Press.
MacLean, P.D. (1996) Paul D. MacLean. In: L.R. Squire (ed.), *The History of Neuroscience in Autobiography, Volume 2*. New York: Academic Press.

Further reading

Corey, G.A. and Gardner, R. (2002) *The Evolutionary Neuroethology of Paul MacLean*. Westport: Praeger Publishers.
Lambert, K.G. (2003) The life and career of Paul MacLean: A journey toward neurobiological and social harmony. *Physiology and Behavior*, 79, 343–349.
Lautin, A. (2001) *The Limbic Brain*. New York: Kluwer Academic/Plenum.
Newman, J.D. and Harris, J.C. (2009) The scientific contributions of Paul D. MacLean. *The Journal of Nervous and Mental Disease*, 197, 3–5.
Papez, J.W. (1937) A proposed mechanism of emotion. *Archives of Neurology and Psychiatry*, 38, 725–743.
Pessoa, L. and Hof, P.R. (2015) From Paul Broca's great limbic lobe to the limbic system. *Journal of Comparative Neurology*, 523, 2495–2500.
Roxo, M.R., Franceschini, P.R., Zubaran, C., Kleber, F.D. and Sander, J.W. (2011) The limbic system conception and its historical evolution. *The Scientific World Journal*, 11, 2427–2440.

Chapter 15

John Carew Eccles (1903–1997)

Eccles is best known for being an outspoken critic of chemical neurotransmission in the CNS – that is, until his own work examining the electrophysiological properties of inhibitory spinal neurons in the 1950s provided compelling evidence for its existence.

John (Jack) Eccles was a native of Melbourne, Australia, whose parents were both school teachers. During his teenage years, Eccles excelled in mathematics (he won a state geometry prize) and athletics (Eccles was a record-breaking Australian universities pole-vaulter) but increasingly became enthused with what philosophers call the mind-body problem – namely, the question of how a physical brain can give rise to consciousness. Believing the answer must lie with a greater understanding of the *synapse* (he would change his mind later in his life), Eccles decided to follow medicine after winning a scholarship to study at the University of Melbourne. By the time he had obtained his MD with first-class honours in 1925, Eccles had already been bestowed one of the world's prestigious academic awards – the Rhodes Scholarship, which allowed him to study at the University of Oxford. Here, Eccles studied physiology and biochemistry for his bachelor's degree and improved his English through interactions with J.R.R. Tolkien. But Eccles had a far loftier ambition for wishing to study at Oxford – he wanted to work under the distinguished neurophysiologist Charles Sherrington who was now in his seventies and still busy studying reflexes in the spinal cord and other areas of the body. As an undergraduate student, Eccles had been strongly influenced by Sherrington's book *The Integrative Action of the Nervous System* and would later describe its author as "the one man in the world whom I wished to have as my master". Although he had never undertaken an experiment prior to his arrival in Oxford, Eccles began his PhD in Sherrington's laboratory with Stephen Creed in 1927 and complete it just 2 years later. This would also prove to be a productive time for Eccles who would publish nearly fifty papers over the next 10 years or so while working at Oxford with a large number of collaborators including the future Nobel laureate Ragnar Granit.

In 1928, Eccles began working with Sherrington in a collaboration that would last for a period of around 3 years. The main focus of this research concerned the neural basis of hind-limb flexor withdrawal in the dog – a response that occurs when the animal removes its back leg from a painful agent such as a pinprick. The main aim was to trace the functional anatomy of this reflex through the spinal cord, which in this case was complex since it involved the flexor muscles of the ankle, knee and hip. Nonetheless, providing the painful stimulus was intense enough, the reflex was very rapid and reliable. It was readily apparent that this reflex not only required the flexor muscles in the withdrawing limb to contract but also the extensor muscles of the opposite limb to relax. In short, the reflex involved different pathways that either had an excitatory or inhibitory effect on their respective muscles.

The concept of inhibition, however, was a poorly understood phenomenon. Sherrington had long recognised that inhibition was part of the normal reflex process, which occurred alongside with excitation. He had also concluded that both excitation and inhibition resulted from activity at the synapse – a site he had named in 1897 to describe the functional contact between nerve cells. In other words, he believed that there were two different types of synapses – excitatory and inhibitory – and both acted on the motor neurons controlling the muscle response. However, not everybody agreed with the idea of an inhibitory synapse. Clearly, an inhibitory state did not travel down the axon, and for some physiologists, this type of activity could be explained through something known as Wedensky inhibition – the idea that inhibition was nothing more than a brief refractory period when the synapse was no longer sensitive to excitation. The answer to this issue was also dependent on a far bigger question – namely, whether synaptic transmission was electrical or chemical. Sherrington had not committed himself to either side, although by 1925 he had conceded that it was feasible that chemical neurotransmission was somehow involved at the synapse. But as Eccles undertook his PhD that examined the time courses of excitatory and inhibitory responses after single volleys of nerve stimulation, he began to come round to a different point of view.

Eccles's views on this matter were also influenced by his work with other investigators at Oxford. This included research examining the effects of stimulating the vagal nerve to the heart and recording the activity of fibres arising from the sympathetic cervical ganglion (a part of the autonomic nervous system that innervates the head and neck). For Eccles, this work revealed that excitatory synaptic neurotransmission took place very rapidly – often in the region of 1 to 2 msecs, which he believed was much faster than could be achieved by chemical liberation. However, inhibition was clearly a much slower process. This type of response was seen, for example, when the vagus nerve to the heart was stimulated – a procedure causing the heartbeat to slow, but only after a period of 100 to 160 msecs had elapsed. These findings led Eccles to conclude that the initial phase of synaptic neurotransmission was electrical, and this acted to produce a "detonator response" in the postsynaptic cell causing it to discharge. But where the synaptic response was much slower, Eccles conceded that a chemical might be involved. Indeed, following Otto Loewi's classic demonstration of neurohumoral release in 1925, much evidence had shown that acetylcholine was the chemical released by the vagal nerve. By 1937, Eccles had accepted that chemical neurotransmission was involved in the peripheral nervous systems but bluntly refused to accept that it occurred in the CNS, which depended on the rapid flow of information.

As Eccles made his views more widely known, he soon became embroiled in an argument, now known as the *sparks and soup controversy*, which was most commonly played out in the regular meetings of the British Physiological Society during the 1930s – although the dispute continued well into the 1950s. The main protagonists of this heated debate were Henry Dale and his pharmacological colleagues who favoured the chemical hypothesis of synaptic neurotransmission, and Eccles along with an array of eminent neurophysiologists who stoutly defended the electrical position. For those looking on, the conflict in these meetings could appear to be rancorous and confrontational. Indeed, Bernard Katz who had just fled from Nazi Germany to England experienced one such meeting in Cambridge in 1935 and wrote: "To my great astonishment I witnessed what seemed almost a stand-up fight between

J.C. Eccles and H.H. Dale, with the chairman E.D. Adrian acting as a most uncomfortable and reluctant referee". But appearances were also deceptive for Dale and Eccles, despite their different styles, were on good terms and often exchanged letters and information. The rivalry would also provide a great incentive for both camps to concentrate their own efforts on trying to prove their respective positions – a situation that was to ultimately prove positive in the long run.

In 1935, Sherrington who had won a Nobel Prize in 1932, retired at the age of 77. Eccles was one of the contenders to succeed Sherrington, but when this did not happen, and with World War II looming, he returned to Australia in 1937 to take up a directorship of the Kanematsu Institute that operated as a diagnostic pathology laboratory for a Sydney hospital. It was an unlikely setting for a research laboratory since the institute was primarily concerned with the routine typing of blood samples and diagnosis of tissue specimens. But Eccles was able to build a new suite of laboratories and attract a number of talented scientists to join him, including Stephen Kuffler and Bernard Katz. For his own part, Eccles turned his attention to the neuromuscular junction – the synapse between motor neuron and muscle end plate. Within a year, Eccles had achieved a major breakthrough by showing that nerve impulses caused tiny voltage increases to occur within the muscle called "end plate potentials" (EPPs). These, in turn, led to the discharge of muscle fibres and contraction. Many researchers by this time had implicated acetylcholine as the transmitter at this peripheral junction – although the rapidity of the muscle response to stimulation again persuaded Eccles that chemical liberation was not involved. But he was to change in mind on this matter in 1942 when Katz and Kuffler found that eserine (a cholinesterase inhibitor) significantly enhanced the effects of acetylcholine at the neuromuscular junction. Eccles was not slow to send an air mail letter to Dale to inform him of his change of mind, although he was still not prepared to give up his cherished electrical synaptic theory with regards to the CNS.

In 1943, Eccles moved to the University of Otago, Dunedin, New Zealand. He was still convinced that chemical transmission only played a minor role at the synapse in the CNS and backed up his claims with several ideas, including the possibility of electrical wave interference and eddy currents as means of inhibiting synaptic activity. Eccles also published an electrical theory of central inhibition in the reputable journal *Nature* in 1945. But the first years of the 1950s would see major new changes in neurophysiological research, with investigators such as Hodgkin and Huxley performing the first intracellular recording from nerve cells (a feat made possible by the discovery of the giant squid neuron by J.Z. Young in 1936). This was also a development in which Eccles wanted to be an active participant, and he set about gaining the technical expertise by recruiting two younger colleagues at Dunedin, Jack Coombs and Lawrence Brock, who between them were able to construct extremely fine glass micropipette electrodes that had a tip diameter of about 0.5 microns (or about fifty-thousandths of an inch). These could then be inserted into spinal cord motor neurons and their activity measured by an amplifier and oscilloscope.

Armed with his new technique, in 1951, Eccles began examining the responses of motor neurons passing out to the quadriceps muscle of the thigh in the anaesthetised cat. When Eccles stimulated the presynaptic neurons with a small electrical current, it failed to trigger an action potential in the motor neuron. However, it did cause the resting potential inside the cell to temporarily increase its voltage slightly – an effect that Eccles called an *excitatory postsynaptic potential* (EPSP). And with increased stimulation, Eccles not only found that the EPSP became more positive, but if raised by enough millivolts, the summation of the EPSPs would trigger a nerve impulse. Today, it is an effect understood by any student who cares to read a textbook on neuroscience.

A year later, in 1952, Eccles used the same procedure to implant a microelectrode into a motor neuron projecting to the quadriceps muscle – but this time stimulated another pathway that came from the hamstrings of the posterior thigh – that is, the opposing muscle to the quadriceps. It was known that these two muscles worked in a reciprocal fashion, so that when one contracted, the other relaxed. Eccles assumed this effect occurred because an electrical excitatory input was being blocked from reaching the output neuron. However, when he stimulated the nerve from the hamstrings, he observed an *inhibitory postsynaptic potential* in the quadriceps motor neuron (i.e. the voltage inside the cell was made even more negative). There was only one reasonable explanation: both excitatory and inhibitory synapses had to exist between the sensory nerves and quadriceps motor neuron, and involve chemical neurotransmission. It was a pivotal moment when Eccles was forced to admit he had been wrong regarding the existence of electrical chemical neurotransmission. There has perhaps never been a greater volte-face in the history of science and Dale, for one, could not resist commenting "a remarkable conversion" had occurred, which was similar to "Saul on his way to Damascus when sudden light shone and the scales fell from his eyes". Yet, while history will tell us that Eccles was wrong for rejecting chemical neurotransmission, his crucial experiment had proven its existence.

In 1951, Eccles took up the chair of physiology at the newly built Australian National University in Canberra, where in his own words he enjoyed "fourteen golden years, scientifically speaking". This period also saw him win a Nobel Prize in 1963. From 1966, Eccles worked in North America, until he was 72 when he retired and moved to Switzerland. Here, he returned to the philosophical issues of the mind-body problem. Interestingly, Eccles admitted that he had long been a committed dualist who had rejected the "irreligious monist materialistic view", which most of his fellow scientists held. In attempting to explain how mind and body interacted, Eccles proposed that it took place at the microscopic level and involved the interaction of non-physical mental units (which he called *psychons*) with cerebral units formed by sets of dendrites (*dendrons*) that used quantum mechanics to affect each other. Eccles was supported in his views by the philosopher Karl Popper, and together they would write *The Self and Its Brain* in 1977. It was described as a searching enquiry by "an agnostic philosopher" (Popper) and a "believer in God and the supernatural" (Eccles). Although they were not in total agreement, Eccles and Popper advocated a pluralistic theory of mind, which held that nature is divided into three different ontological domains (or worlds) in reciprocal interaction. These domains can be (very loosely) called: the material, the mental and the world of knowledge. It may come as a surprise to some readers that nearly 20% of Eccles written output, which consists of 568 publications, concerns these broader philosophical issues. Eccles died in 1997 in his home in Contra, Switzerland, at the age of 97 and was survived by his nine children.

Eccles's major writings

Brock, L.G., Coombs, J.S. and Eccles, J.C. (1952) The recording of potentials from motor neurons with an intracellular electrode. *Journal of Physiology (London)*, 117, 431–460.

Eccles, J.C. (1953) *The Neurophysiological Basis of Mind*. Oxford: Clarendon Press.

Eccles, J.C. (1957) *The Physiology of Nerve Cells*. Baltimore: John Hopkins Press.
Eccles, J.C. (1961) The Ferrier Lecture: The nature of central inhibition. *Proceedings of the Royal Society of London, Series B*, 153, 445–476.
Eccles, J.C. (1964) *The Physiology of Synapses*. New York: Academic Press.
Eccles, J.C. (1977) My scientific odyssey. *Annual Review of Physiology*, 39, 1–13.
Eccles, J.C. (1982) The synapse: From electrical to chemical transmission. *Annual Review of Neuroscience*, 5, 325–339.
Eccles, J.C. (1992) Under the spell of the synapse. In: F.G. Worden, J.P. Swazey and G. Adelman (eds), *The Neurosciences: Paths of Discovery, 1*. Berlin: Springer.
Popper, K.R. and Eccles, J.C. (1977) *The Self and its Brain*. Berlin: Springer International.

Further reading

Burke, R.E. (2006) John Eccles' pioneering role in understanding central synaptic transmission. *Progress in Neurobiology*, 78, 173–188.
Cavanna, A.E. and Nani, A. (2014) John Eccles and Karl Popper. In: *Consciousness: Theories in Neuroscience and Philosophy of Mind*. Berlin: Springer.
Curtis, D.R. and Andersen, P. (2001) John Carew Eccles 1903–1997. *Historical Record of Australian Science*, 13, 439–473.
McComas, A.J. (2011) *Galvani's Spark: The Story of the Nerve Impulse*. Oxford: Oxford University Press.
Shepherd, G.M. (2010) *Creating Modern Neuroscience: The Revolutionary 1950s*. Oxford: Oxford University Press.
Valenstein, E.S. (2005) *The War of the Soups and the Sparks*. New York: Columbia University Press.

Chapter 16

Bernard Katz (1911–2003)

Bernard Katz laid the foundations for our understanding of almost every aspect of synaptic transmission – from recognising the importance of calcium ions in presynaptic physiology to establishing the quantal theory of neurotransmitter release.

Katz was born in Leipzig to a Russian fur trader who never bothered to apply for citizenship in Germany and a Jewish Polish mother. This not only meant that they were all officially regarded as "enemy aliens" during World War I (1914–1918) but also that Katz grew up stateless. Nonetheless, despite the widespread anti-Semitic prejudices of the time, which led Katz to consider emigrating to Palestine, he did well at school and chose to study Greek and Latin over mathematics because it gave him more time to play chess. In 1929, Katz managed to find a place at Leipzig University where he began studying medicine – a profession that offered him a good chance of earning a living outside Germany. This would also lead him to work under Martin Gildemeister who was well known for his mathematical approach to physiology. Completing his state examinations in 1934, Katz wrote his MD thesis the same year based on research examining the membrane permeability of frog muscle. This work would lead to two papers being published in the highly respected *Pflügers Archives* – the oldest physiological journal in the world founded in 1838. These papers turned out to be crucially important for Katz since they impressed Archibald Hill at University College, London. A leading light of British science and culture, and a strong critic of Hitler, who worked tirelessly to help colleagues disaffected by his tyranny, Hill was happy to invite Katz to his biophysics laboratory after receiving references from Gildemeister and Chaim Weizmann who was later to become the first president of Israel. After a third-class train journey to Holland and a sailing on the Flushing-Harwich ferry, Katz arrived in London in 1935 possessing no more than a League of Nations stateless certificate[1] and just £4. Initially supported by a refugees fund, Katz obtained his PhD just 3 years later after examining the neurophysiological properties of the myelinated frog sciatic nerve.

During his time at University College, Katz lived with Hill in London, in effect becoming part of his family. Katz also spent some time at the Plymouth Marine Biological laboratory where he worked with Alan Hodgkin and Andrew Huxley. Another important contact was John Eccles who would move back to Australia in 1937 and invite Katz to work alongside him with Stephen Kuffler. Unfortunately, when Katz reached Australia in 1939, World War II had just broken out and he would spend 9 months serving as a radar operator behind Japanese lines in New Guinea for the Australian Air Force. He also served time as a flight lieutenant. After the war, in 1946, Katz received a telegram from Hill inviting him back to London to take up assistant directorship of his biophysics department. Katz remained here for the rest of his working life. Not long after this, in 1948, a young American called Paul Fatt with a bachelor's degree in biochemistry, who had also seen action in the Normandy Landings, began his PhD with Katz, examining synaptic transmission at the frog neuromuscular junction.

This was a period in neuroscience when there was still controversy over the nature of chemical neurotransmission, and the neuromuscular junction provided a good model system by which to understand the phenomenon. Fatt also joined Katz's laboratory at a propitious time, since the intracellular recording microelectrode, made from thin glass pipettes that are pulled to a very fine and sharp tip of some 50 to 500 nanometres (a nanometre is one billionth of a metre), had just been invented in Chicago by Gilbert Ling and Ralph Gerald in 1949. Up to this point, previous work had been carried out by external recording electrodes from bundles of muscles fibres – much of it by Eccles. But armed with the new microelectrode, Katz and Fatt were able to implant its tip into the end plate of the frog's toe muscle and then stimulate its motor neuron with single pulses of electrical activity. The result was a brief and small rise of voltage in the muscle fibres that were known as the end plate potential (EPP). Although the size of the EPP was small (e.g. around 5 or 10 mV), it was normally enough to make the muscle contract. This effect was abolished, however, if the cholinergic blocker curare was applied to the end plate. In contrast, muscle excitability was increased by the acetylcholinesterase inhibitor eserine that increased levels of acetylcholine in the synapse. These results were not unexpected: Katz had obtained similar results while working with Eccles in the 1940s. Nonetheless, they provided yet further support for the existence of cholinergic neurotransmission at the neuromuscular synapse – work that had begun with Henry Dale and his colleagues in the 1930s.

It was during this research, however, that Katz and Fatt noticed something new. While recording the EPPs from the muscle fibres, their cathode ray screen sometimes showed dancing "blips" of extremely small voltages (e.g. around 0.5 mV) at random intervals in the absence of any motor neuron stimulation. Because these blips were so tiny, Katz and Fatt called them miniature end plate potentials (mEPPs). In fact, Katz had seen these responses some 2 years previously when working on the muscle spindle of the frog but dismissed them as an artefact derived from the electrical "noise" of his instrumentation. Now, Katz and Fatt decided to examine these tiny blips in more detail by seeing if they still occurred when curare was applied to the end plate – a procedure that they had already shown abolished the much larger EPPs. To their delight, the application of curare eliminated the mEPPs too. There was only one viable conclusion: the "noise" was being produced by biological activity – most likely the transient release of acetylcholine from the nerve endings. This was confirmed when they cut the motor nerves innervating the motor end plate and found this abolished all mEPP activity in the muscle cells.

To assess the nature of these tiny potentials further, Katz and Fatt decided to manipulate the ionic composition of the solution bathing the neuromuscular junction. They started by removing sodium ions (Na+) and found this abolished both the EPPs and mEPPs. This finding was not a surprise since sodium was well established as an ion with electrical properties (as shown by Hodgkin and Huxley) and its influx into the end plate known to be the cause of muscle contraction. However, when Fatt and Katz reduced the levels of calcium (Ca^2+), they found an interesting difference: this manipulation significantly reduced the amplitude of the EPPs (in effect making the muscle less likely to contract), but it had no effect on the occurrence of the much smaller mEPPs. Although it was then speculation, these results indicated that the EPPs were caused by the secretion of acetylcholine from the axon terminals through some form of process involving calcium. In contrast, the spontaneous mEPPs were tiny random events brought about by small packages of acetylcholine "escaping" from the presynaptic motor neuron independent of calcium.

One of the key features of the mEPPs was they were highly consistent electrical events – each one being in the region of 0.5 mV. This characteristic turned out to be an important one; on closer inspection, Katz and Fatt realised that each mEPP was actually an integral multiple (0, 1, 2, 3 or more) of the much larger EPP. This was a highly informative discovery since Fatt and Katz could now deduce that the EPP that caused a muscle to twitch was the sum result of very small packages (or what they called "quanta") of acetylcholine the size of a mEPP. Or put another way: acetylcholine was not being released in a continuous secretary flow from the presynaptic ending but in the form of small units, or packages, of fixed size.

In the summer of 1952, Fatt completed his PhD and moved to Australia where he worked with John Eccles. Katz now began to collaborate with the Spaniard José del Castillo who had a lecturing post at London's University College. One question that arose from the quantal theory was: how much acetylcholine is stored in a quantal package? Since the amounts were far too small to be measured directly, Katz and del Castillo addressed this problem by using a specialised form of statistical inference known as Poisson analysis. Their calculations indicated that a large number of acetylcholine molecules were released in one quanta – perhaps as many as 10,000. They also estimated about 200 quanta were needed to create a typical EPP. Shortly after this analysis, the first detailed pictures of the neuromuscular junction were produced by the electron microscope in 1954.[2] It lent further support to the theory of quantal release for the pictures revealed the axon terminals contained many small "vesicles" that provided the repositories for storing neurotransmitter molecules. It also became apparent that neurotransmitter release occurred when these vesicles fused with the membrane in a process known as exocytosis, causing their contents to be released into the synapse.

The work by Katz and Fatt in the early 1950s, followed by the findings of Katz and del Castillo, were pivotal moments in the history of synaptic physiology because they established that neurotransmitters are released in individual packages of "quanta". Thus, at any given synapse, the amount of neurotransmitter release is always in fixed quantities and never less than a certain amount. It is also apparent that an action potential arriving at the presynaptic terminal is able to cause an EPP in the muscle because it synchronises the release of many transmitter quanta. This fundamental insight into the process of synaptic neurotransmission led to Katz winning a Nobel Prize for his answering "questions of fundamental importance for the understanding of the mechanism underlying the transmission between nerve cells" in 1970 (along with Julius Axelrod and Ulf von Euler).

Another scientist to work with Katz was the Mexican Ricardo Miledi, an ex-medical student who had learned how to dissect single nerve fibres and pull glass pipettes by hand – skills that had won him scholarships to work at Woods Hole, Massachusetts, with Stephen Kuffler, and then in Australia with John Eccles. Invited to join Katz's laboratory in the late 1950s, Miledi set about establishing the precise role of $Ca^{2}+$ in synaptic neurotransmission. Although Katz and Fatt had clearly shown that $Ca^{2}+$ was involved in the release of acetylcholine – how it did so was unclear. Working with the frog neuromuscular junction, Miledi and Katz removed $Ca^{2}+$ from the bathing solution, and found that the nerve impulse still entered the presynaptic motor nerve terminal – although it now did not cause the secretion of any neurotransmitter. However, just a small amount of $Ca^{2}+$ added back to the bathing fluid restored the transmitter release. The importance of $Ca^{2}+$ was further emphasised when Miledi and Katz were able to block the flow of the nerve impulse by stopping the influx of Na+ with tetrodotoxin (a poison derived from the Japanese puffer fish). Again, providing there was some $Ca^{2}+$ in the bathing fluid, a brief depolarising pulse applied to the ending of a motor nerve caused packets of acetylcholine to be released. In short, neurotransmitter release did not require Na+ or an action potential – only presynaptic depolarisation.

The most likely explanation was that the presynaptic depolarisation was being caused by $Ca^{2}+$ that entered the presynaptic terminals through specialised voltage-gated channels. From 1965 to about 1969, Miledi and Katz spent part of their summers working in Naples, Italy, to investigate this issue using the giant synapse of the squid – a preparation that is large enough to allow a direct injection of $Ca^{2}+$ into the presynaptic synaptic region. This work would finally prove (at least beyond reasonable doubt) that transmitter release is brought about by an influx of external calcium through special membrane channels which are opened by the depolarising pulse (Katz and Miledi 1970). Although there was still much to clarify Katz and Miledi had outlined a basic process that is now recognised as an essential feature of all neurotransmitter release.

In his later years, Katz began to unravel the biochemistry of the pineal gland and investigate how melatonin is produced in response to light. He would continue to be head of the biophysics department (a post he had taken over from Hill in 1952) until his retirement in 1978. Although he was succeeded by Miledi (who left for California in 1983), Katz still worked on a regular basis as emeritus professor. He also remained active on research council committees and other types of science administration. By this time, he had won many awards in addition to his Nobel Prize. This included being elected a fellow of the Royal Society in 1952 (he would win its coveted Copley Medal in 1967), serving as its vice president on two occasions and a knighthood in 1969. His work was also recognised in his birthplace of Germany with a Pour le mérite in 1982 despite never having been a German citizen. Although English was not his first language, his writings were recognised as having a prose that was simple, straightforward and precise. He was also said never to use a long word when a simpler one would do. An intense and private man with a wry smile, who played fiercely at chess, Katz would be active throughout his eighties. He became hospitalised in 2003 and died later that year at the age of 92.

Notes

1 These were known as a Nansenpass – named after the Norwegian Fridtjof Nansen (1861–1930) who won a Nobel Peace Prize in 1922 for his efforts to help war refugees. He actually began his academic career as a marine biologist, who made an important contribution to establishing the neuron doctrine with his Golgi staining of nervous material.

2 The assignment of priority for this discovery is contentious since two independent groups (De Robertis and Bennet in Seattle and George Palade and Sanford Palay in New York) published their findings at around the same time.

Katz's major writings

Fatt, P. and Katz, B. (1951) An analysis of the end-plate potential recorded with an intracellular electrode. *Journal of Physiology*, 115, 320–370.

Del Castillo, J. and Katz, B. (1954) Quantal components of the end-plate potential. *Journal of Physiology*, 124, 560–573.

Katz, B. (1962) The Croonian Lecture. The transmission of impulses from nerve to muscle, and the subcellular unit of synaptic action. *Proceedings of the Royal Society of London, Series B, Biological Aspects*, 155, 455–477.

Katz, B. (1966) *Nerve, Muscle and Synapse*. New York: McGraw Hill.

Katz, B. (1969) *The Release of Neural Transmitter Substances*. Liverpool: Liverpool University Press.

Katz, B. (1970) On the quantal mechanism of neural transmitter release. In: *Nobel Lectures, Physiology or Medicine 1963–1970*. Amsterdam: Elsevier (1972).

Katz, B. (1996) Neural transmitter release: From quantal secretion to exocytosis and beyond: The Fenn Lecture. *Journal of Neurocytology*, 25, 677–686.

Further reading

Augustine, G.J. and Kasai, H. (2007) Bernard Katz, quantal transmitter release and the foundation of presynaptic physiology. *Journal of Physiology*, 578.3, 623–625.

Bennet, M.R. (2001) *History of the Synapse*. Amsterdam: Harwood Academic Publishers.

Bynum, W.F. (1976) A short history of the physiological society (1926–1976). *Journal of Physiology*, 263, 23–72.

Colquhoun, D. (2003) Bernard Katz 1911–2003. *Physiological News*, 52, 34–38.

Jeng, J.M. (2002) Ricardo Miledi and the calcium hypothesis of neurotransmitter release. *Nature Reviews Neuroscience*, 3, 71–76.

Sakmann, B. (2007) Bernard Katz. *Biographical Memoirs of the fellows of the Royal Society*, 53, 185–202.

Chapter 17

Seymour Solomon Kety (1915–2000)

Kety is best remembered for inventing a method for measuring cerebral blood flow, which helped lay the foundations for the development of functional brain scanning techniques and providing the first definitive evidence for the genetic basis of schizophrenia.

Seymour Kety was born in Philadelphia to Jewish parents who despite having little money managed to raise their son in an intellectually stimulating home, surrounded by a good supply of books, great music and lively discussion. At the age of 7, Kety badly fractured his leg in a car accident, causing an injury that required a long stay in hospital and leaving him with one leg shorter than the other. This forced hospitalisation, nevertheless, fostered in the young boy a love of reading, especially after his family bought him the encyclopaedic *Book of Knowledge* – a series of twelve volumes that covered a large panorama of disciplines including science, geography, philosophy and literature. Kety read the entire set of books – a pursuit that more than made up for his disrupted schooling. Some 3 years later, at the age of 10, Kety was given a chemistry set by an aunt. This quickly led to an obsession, with Kety building a laboratory at home and saving his school lunch money so he could buy chemicals. He would even write a book at this tender age (not published) entitled *The Boy's Book of Chemistry*, which described the principles of chemistry through various historic or exciting experiments. By the time he went to high school in 1928, he was presumably proficient in chemistry. Kety attended Central High in Philadelphia, the city's premier high school that has the distinction of being the only one in the United States to confer academic degrees on its graduates (the school had professors – not teachers). Here, Kety excelled in physics[1] and enjoyed Greek, Latin and literature. Leaving school in 1932, Kety entered the University of Pennsylvania where he graduated Phi Beta Kappa (an award only conferred on the most distinguished students) in 1936. He then won a scholarship to study medicine, although was soon contemplating research after finding a summer job with a biochemist working for the lead industry. This required Kety to analyse urine samples of lead workers – a task leading him to realise that sodium citrate might be beneficial in those with its poisoning. It was a theory later tested by giving rats lead-enriched diets as a medical student, leading to his first published paper in 1942. Obtaining his MD in 1940, Kety began an internship in paediatrics (the speciality of his wife), followed by a grant to work with Joseph Aub, an expert on lead poisoning at Boston's Massachusetts General Hospital. However, after arriving in Boston, Kety found Aub had changed his research interests to traumatic and haemorrhagic shock – an area more suited to wartime concerns. Undeterred, Kety joined Aub's group and soon found the physiology of circulation to be a fascinating new area of interest.

In 1943, Kety returned to the University of Pennsylvania to teach pharmacology and join forces with Carl Schmidt, then a leading figure in the field of blood circulation. Kety was keen to follow up on a discovery made by Aub's group, which had shown that in cases of traumatic shock, the body reflexively redistributes cardiac output to the brain, heart and lungs because these organs can only survive a short time without blood flow. At the time, Schmidt was working on a related problem by attempting to measure the cerebral blood flow (CBF) in the anaesthetised rhesus monkey by means of a bubble flow meter inserted in the cerebral arterial system. This device enabled the calculation of air and gas velocities entering the brain, thereby providing an indirect judgement of CBF. Inspired by this work, Kety began to wonder if there was a way of measuring CBF, which could be used in humans without surgical invasion. He was well aware that if such a method could be discovered, then it would also be useful in exploring the functions of the brain, along with the diagnosis of neurological or mental disorders.

One possible solution to the problem was to make use of a phenomenon known as The Fick Principle (named after Alfred Fick in 1870), which shows that blood flow to an organ can be calculated, provided one finds a marker substance that is taken up, or utilised, by the organ. If this criterion is met, then one can subtract the concentration of that marker substance entering an organ in the arterial supply from that leaving in the venous blood to get a measure of blood flow. In fact, this type of method had been successfully used to assess blood flow in the kidney and liver, since both these organs excrete foreign substances into the circulation at a rate that can be measured. Similarly, cardiac output had been calculated by determining the oxygen in the blood entering and leaving the lungs. Unfortunately, for Kety, there was no obvious substance that the brain removed from or excreted into its circulation.

This state of affairs changed in 1943 when Kety hit upon a solution that entailed introducing an inert gas into the blood, which the brain could take up by simple diffusion. The gas Kety chose was nitrous oxide (N_2O), which had the benefit of being inhalable with air. He began his first preliminary studies on monkeys and found that after about 10 minutes of inhalation, the blood and brain concentrations of N_2O reached equilibrium. But more importantly, the rates at which this occurred in the arterial and venous blood was different – and it was this disparity that enabled Kety through some clever calculations to arrive at a measure of blood flow, along with some estimates of oxygen consumption. The results of the Kety-Schmidt technique, as it became known, on human volunteers was first published in 1948 in five separate articles in a single issue of *Journal of Clinical Investigation*. It was a technique that was to start a revolution in neuroscience.

This blood flow analysis produced some remarkable findings. For example, it revealed that the human brain that only weighs about three pounds (about 2% of total body weight) accounted for some quarter of the body's blood flow and energy metabolism. Kety's work also answered a long-standing question that had baffled physiolo-gists for over a half a century. This was whether alterations in flow of blood entering the brain were due to extrinsic bodily factors, such as changes in systemic blood pressure or cardiac input, or whether it was an adaptation undertaken by the brain itself.[1] The results were unequivocal: Kety showed that the brain contained mechanisms that regulated its own flow. Indeed, we now know that when neurons use more oxygen, chemical signals cause nearby blood vessels to dilate, which in turn leads to a local increase in the flow of blood.

Kety also used his new technique to measure blood flow in anaesthesia and sleep, along with various medical conditions such as hypertension, epilepsy and senile dementia. The most striking results were found with deep anaesthesia where the cerebral oxygen consumption fell by 50%. But overall, increased flow or metabolism in all other conditions was found to be small or negligible. Of particular interest to Kety was schizophrenia. To his surprise, he found that schizophrenics exhibited normal blood flow – although he suspected this was because the illness only affected small localised regions of the brain, which his rather gross method was unable to detect. Nonetheless, this problem encouraged Kety to think about how he might go about measuring localised activity, and in the mid-1950s, he came up with a method for use in experimental animals, which involved injecting a compound called "trifler-iodomethane" into the blood, which was tagged with a radioactive form of iodine. This radioactive tracer then showed up in slices of frozen brain tissue. Kety first used this method in 1955 with cats by presenting light flashes to the retina – a procedure that showed increased activity in the visual cortex. Although simple, it can be regarded as the first ever example of functional brain imaging, and Kety's method would later be resurrected for human use in the 1970s – but this time with radioactive water as the tracer and positron emission tomography (PET) replacing the autoradiography.

In 1950, Kety was offered the post of first scientific director of the newly established NIMH in Bethesda – which was to soon become the largest organisation in the world specialising in mental illness research. Here, Kety's remit was to select a group of scientists who were required to produce groundbreaking discoveries. He would not fail for several of his group would go on to win the Nobel Prize and many others elected to the illustrious National Academy of Sciences. One of the scientists was Louis Sokoloff, who like Kety came from Philadelphia and was a MD graduate from the University of Pennsylvania. Invited to work at the NIMH in 1953, Sokoloff also became involved in measuring brain blood flow and metabolism. This would lead to another major breakthrough when, in the 1970s, Sokoloff developed a more powerful way of measuring brain activity – this time by injecting into animals a radioactive labelled analogue of glucose called "2-deoxyglucose" (2-DG). This method works on the principle that the brain is almost entirely dependent on glucose for its energy, with the consequence that the most active areas of the brain will show the greater accumulation of 2-DG. Soon after Sokoloff's breakthrough with 2-DG, his method would be adapted for use in PET with human subjects.

In 1956, Kety resigned as director of the NIMH to head the Clinical Science Laboratory where he established a program of research into the biology of schizophrenia. In particular, he began to focus on the genetics of this illness. Although there was evidence to show that schizophrenia ran in families, most psychiatrists at the time believed this occurred because the illness was learned from one's parents and family. Consequently, the only viable treatment was some form of psychotherapy. Kety decided to examine this question more astutely by gathering a sample of adopted individuals from Denmark. The Danes are a fairly homogenous group genetically, with good records of their citizens, and this allowed Kety to collect data, over the course of 25 years, on over 14,000 adoptees. His most important finding was that children born to a schizophrenic mother, but taken away and reared in a normal adoptive family, were just as likely to develop psychosis as those reared by the biological mother. In other words, the type of family upbringing had little impact on the development of schizophrenia. Another significant finding was that biological relatives of a child with schizophrenia had a greater incidence of this illness in their families, in contrast to those who had acted as the adoptees. It was convincing evidence to show that genetic inheritance played a major role in schizophrenia – and it pointed towards medical treatment rather than psychotherapy as the more effective remedy. This was very much in the spirit of the times as the first antipsychotic drugs such as chlorpromazine were just beginning to be developed and marketed.

Kety also looked at the borderline symptoms of schizophrenia that were deemed to be much less severe forms of the illness. He found that many of these symptoms, such as poor interpersonal skills and impaired emotional responding, were often present in the relatives of the schizophrenic but not in the adoptive family members. Thus, Kety came to recognise that there is a latent form of schizophrenia which is still strongly genetic but which produces a less debilitating form of the illness. This finding was the beginning of the idea of the "schizophrenic spectrum" that presumes there are several conditions that don't fully qualify as schizophrenia, yet they are related to the psychotic condition, and they increase one's vulnerability to developing the illness. Today, the term "schizotypical personality disorder" is often used in this context.

In 1967, Kety moved to Harvard University, becoming director of the psychiatric research laboratory at Massachusetts's General

Hospital, although he returned to the NIMH as an emeritus scientist in 1981. Kety's legacy to neuroscience encompasses at least three major areas of advancement. First, as a physiologist he was the first to measure CBF in the human brain, a breakthrough that led directly to the development of functional brain imaging techniques such as PET. Second, as a leader and administrator, he established the NIMH as a world-renowned centre of research with a plethora of leading scientists. And third, despite having never trained as a psychiatrist, he clearly changed the course of psychiatry by placing it on a stronger biological foundation. But what is perhaps even more remarkable about the man is the affection in which he was held by his many students and co-workers. Despite the many prestigious honours and awards bestowed on him in his later years, according to Louis Sokoloff (Kety) "remained the same humble, modest, self-effacing, unselfish, considerate, generous and warm human being that he was 56 years ago when I first met him". And another much loved personality trait was his enormous reservoir of jokes and amusing anecdotes, which he sometimes used to great effect in order to make a point. For someone who had spent much of his career holding an executive position over many of his colleagues, to have such adulation is some accomplishment. Kety died in 2000 at the age of 84.

Notes

1 Kety recounts in an autobiography that as a student he wrote to Albert Einstein and was immensely gratified to receive a splendid letter by the great man in reply.
2 This question went back to 1890 when Charles S. Roy and Charles Sherrington first attempted to measure blood flow in an anaesthetised dog by using a device placed on the surface of its brain.

Kety's major writings

Kety, S.S. and Schmidt, C.F. (1948) Nitrous oxide method for the quantitative determination of cerebral blood flow in man: Theory, procedure and normal values. *Journal of Clinical Investigation*, 27, 476–483.

Kety, S.S. (1950) Circulation and metabolism of the human brain in health and disease. *American Journal of Medicine*, 8, 205–217.

Kety, S.S. (1961) A biologist examines the mind and behavior. Many disciplines contribute to understanding human behaviour each with peculiar virtues and limitations. *Science*, 132, 1861–1870.

Kety, S.S., Rosenthal, D., Wender, P.H., Schulsinger, F. and Jacobsen, B. (1976) Mental illness in the biological and adoptive families of adopted individuals who have become schizophrenic. *Behavioral Genetics*, 6, 219–225.

Kety, S.S. (1979) The metamorphosis of a psychologist. *Annual Review of Neuroscience*, 2, 1–15.

Kety, S.S. (1996) Seymour S. Kety. In: L.R. Squire (ed.), *The History of Neuroscience in Autobiography. Volume 1*. Washington, DC: Society for Neuroscience.

Further reading

Holtzman, P.S. (2001) Seymour S. Kety and the genetics of schizophrenia. *Neuropsychopharamcology*, 25, 299–304.

Portnow, L.H., Vaillancourt, D.E. and Okun, M.S. (2013) The history of cerebral PET scanning: From physiology to cutting-edge technology. *Neurology*, 80, 952–956.

Small, S.S. (2004) Quantifying cerebral blood flow: regional regulation with global implications. *The Journal of Clinical Investigation*, 114, 1046–1048.

Sokoloff, L. (2001) Memorial for Seymour S. Kety. *Neuropsychopharmacology*, 25, 291–293.

Sokoloff, L. (2003) Seymour S. Kety 1915 – 2000. In: *Biographical Memoirs of the National Academy of Sciences. Volume 83*. Washington, DC: The National Academy Press.

Traystman, R.J. (2004) The paper that completely altered our thinking about cerebral blood flow measurement. *Journal of Applied Physiology*, 97, 1601–1602.

Chapter 18

Brenda Milner (b. 1918)

Milner is most famous for describing the severely amnesiac patient Henry Molaison, which revealed the importance of the hippocampus in long-term memory, although she has also carried out pioneering work on the frontal lobes and hemispheric specialisation, as well as being credited with being a founder of cognitive neuroscience.

When Brenda Milner (née Langford) was born in Manchester, England, in 1918, there was little to suggest she would become a pioneering explorer of brain function and world-renowned neuroscientist. Her father was a music critic for the *Manchester Guardian* who gave piano lessons in his spare time and her mother an office clerk from a broken home who left school at the age of 14. Milner's father also took to tutoring his daughter rather than sending her to school, with the result that she later recalled her home was filled with music, books on art and many of the classics including Shakespeare. She also began to learn German and French from her parents at a young age. But there was little in the household for the young girl to read on science. At the age of 6, Milner and her mother contracted influenza – most likely encephalitis lethargica – of which there was an outbreak in Britain in 1924. This disease was associated with a high rate of mortality and sometimes led to post-encephalitic Parkinsonism as popularised in the book and movie of *Awakenings*. Fortunately, both mother and daughter recovered. Two years later, her father succumbed to tuberculosis and Milner's education took a more conventional route when she was obliged to join a local girl's school. This was Withington Girls School, which still provides education for girls between the ages of 7 and 18 and has a list of notable alumni. Here, despite her background in literature and languages, Milner became enamoured by "the beautiful logic" of mathematics, an interest leading her to win a place at the all-female Newnham College, Cambridge, in 1936. Her fascination with mathematics would wane, however, after just 1 year when realising she was not going to distinguish herself in the discipline. Looking around for another subject, she considered philosophy but was dissuaded after the senior members of her college warned her it was unlikely she would earn a living from its study. Instead, she plumped for psychology – a subject she knew absolutely nothing about. It was a wise choice for Milner quickly realised her new subject suited her inquisitive temperament, and she graduated with an upper second BA degree in experimental psychology in 1939 – a classification she later admitted she "hated" obtaining. At Cambridge, she also came under the influence of Oliver Zangwill, at the time a research student in the laboratory of memory psychologist F.C. Bartlett, who acted as Milner's supervisor and tutor from 1937 to 1939. According to Milner, he was particularly important in impressing upon her the need to study brain damage as a means of understanding higher cognitive function. Later, in 1952, Zangwill became professor of the psychology department, where he is widely credited with establishing the science of clinical neuropsychology.

After her degree, Milner was keen to pursue postgraduate research in Cambridge, although the outbreak of World War II in 1939 meant this had to be redirected towards the war effort. At first, Milner became part of a team involved in developing selection tests for fighter and bomber pilots. But in 1941, she moved to Christchurch, Dorset, where she was employed evaluating different types of radar display for the military. Although she would only spend 6 months in Christchurch before being relocated to Malvern in 1942, it proved a significant turning point in her life – for it was here she met a young electrical engineer from Leeds University called Peter Milner, who would later become her husband. The small town of Malvern also proved congenial, and Milner spent two and a half pleasant years there, before being called back to Cambridge in mid-1944.

While happy to be back at university, that same year, Peter Milner was invited to join a small group of physicists to work on an atomic energy project for the British Government in Montreal, Canada. Now in a close relationship with Brenda and not wanting to travel without her, the two agreed to a quick registry office marriage. Soon after, the newlyweds boarded the *Queen Elizabeth* in Glasgow with just £5 between them. It was also a perilous journey due to the German U-boats, and their ship had to take a lengthy zigzag course to reach North America. On arrival in Montreal, Peter took up his new post, while Brenda looked for work. She had hoped to join the psychology department at McGill University, only to find the war years had caused it to fall into a dilapidated state. Despite this setback, Brenda got a position teaching psychology, in French, at the nearby Université de Montréal, which was run by Dominican monks. Although Milner had spoken little French since high school, she had read a great deal of French literature and she spent 7 years in her post. But it was McGill that continued to attract her, and in 1947, its psychology department received a boost with the arrival of Donald Hebb, who was also employed by Wilder Penfield at the MNI. Hebb had also just completed his book *The Organization of Behavior* and used the chapters for a series of evening seminars that attracted a devoted group of participants. Inspired by the debates, Milner cautiously approached Hebb in 1949 and "persuaded" him to let her do a PhD under his supervision.

Soon after starting her PhD, in June 1950, Milner was offered the chance of working at the MNI, testing the cognitive effects of temporal lobe ablation. It was an offer she could not refuse since so little was known about the effects of temporal lobe lesions in humans. Because she was armed with little more than a few simple tests bequeathed to her by Hebb, Milner had to rely heavily on animal studies to assess what deficits she was likely to find. One known effect in monkeys, reported by Kluver and Bucy in 1938, was a flattening of emotion along with an odd abnormality known as psychic blindness. A few years later, Mishkin and Pribram showed that inferotemporal (IT) cortex lesions in monkeys impaired performance on visual discrimination. As a result, Milner began testing her patients on tasks requiring them to spot visually incongruous objects in picture scenes – work that was to show perceptual deficits which became more marked after right hemisphere lesions.

By now, Penfield was beginning to be a little more adventurous with his surgery. At first, he had usually confined his lesioning to the most anterior regions of the temporal cortex. This produced little behavioural impairment, although it was not always successful at curbing the seizures, with the result Penfield began extending his lesioning to more medial parts of the temporal lobe, including the amygdala and hippocampus. In some instances, this led to unexpected effects. One case was a civil engineer known as PB. Following a second operation in the early 1950s that invaded his left-sided medial temporal cortex and hippocampus, he experienced severe anterograde amnesia. At first, this was considered a unique case. But in 1952, another patient, FC, suffered a similar deficit after a temporal lobotomy, which included parts of the hippocampus.

In 1955, these two cases were presented to a meeting of neurosurgeons in Chicago. A few days later, Penfield received a call from William Scoville who practised neurosurgery in Hartford, Connecticut. According to some accounts, Scoville was a brash character, known for his reckless driving and high-risk pranks such as climbing the cable tower of the George Washington Bridge at night. At the time, lobotomies were being performed regularly in the United States to treat various forms of mental illness, and Scoville was at the forefront of this "psychosurgical" movement. Scoville told Penfield that he had observed a pronounced memory loss in one of his own patients whom had received a bilateral lesion of the medial temporal lobes for the treatment of his seizures. As a result, Milner was invited to travel to Hartford to study the patient's amnesia.

The patient, of course, was Henry Molaison, known at the time by his initials HM,[1] who is perhaps the most famous patient in the annals of memory research, if not psychology itself. HM was born in 1926 and had a normal childhood until he was involved in a bicycle accident at the age of 7 that left him unconscious for several minutes. This was probably the incident that led to HM having seizures at the age of 10, which worsened progressively through his teenage years and early twenties. Although he worked for a time on an assembly line, by the age of 27, HM had become so incapacitated, a normal life was impossible. As a result, HM was taken to meet Scoville who identified the medial temporal lobes as the origin of his seizures. With the approval of patient and family, Scoville performed surgery, which he undertook in August 1953. He administered a local anaesthetic to the scalp, bored two holes in the front of HM's skull and pushed the frontal lobes up with a spatula to gain access to the medial regions of the temporal lobes. By inserting a knife and suction pump to remove tissue, Scoville removed roughly the size of a tennis ball from each hemisphere. This included most of the hippocampus, amygdala and surrounding cortex.

When Milner first met HM, she saw his seizures were now under control, but his amnesia was far worse than she could have expected. To a casual observer, HM appeared normal. He scored normally on IQ tests and did not suffer any perceptual or temperament problems. Yet, he forgot the daily events of his life, almost immediately as they occurred. This meant he never came to recognise any of the hospital staff or even learn to find his way to the bathroom. He could read the same magazine over and over without becoming familiar with the contents. HM also underestimated his own age, apologised for forgetting the names of persons he had just met and described his state as "like waking from a dream . . . every day is alone in itself". Nor sadly was there ever any improvement. Indeed, after his surgery, HM had moved to a new neighbourhood, and despite spending 10 months living there, he could not recall his address, although he remembered the old one perfectly.

Scoville and Milner published a brief account of HM in 1957, and the paper has since become one of the most cited in the history of neuroscience, with over 2500 citations in major journals. One of the reasons for this interest lay in the fact that hitherto, the hippocampus had almost been overlooked as being involved in memory. A possible reason for this ignorance may have been due to Karl Lashley's influential research that had apparently convinced many that memory functions were widely distributed throughout the cortex and integrated with intellectual and perceptual functions. The case of HM showed this theory to be untenable. But Milner's work with HM, which was to be continued over five decades, was to open the door for much more. It not only established the crucial role of the hippocampus in the consolidation of long-term memory but also showed that multiple memory systems existed in the brain with their own specialised functions. In particular, her work would show the hippocampus to be involved in a system responsible for declarative or conscious recollection of information – but not in procedural or motor-based memory. In fact, Milner demonstrated the intact procedural ability in an early experiment when she found HM was able to learn to draw accurately around a complex shape (a star) using a mirror for visual guidance, which improved over several days, although he never had any conscious recollection of the task.

If these advances into understanding the role of the medial temporal lobes in memory were the only accomplishments of Milner's, then they would have been sufficient for her to be placed into the pantheon of great psychologists. But there would be more to come over the course of her extraordinarily long and productive career. During the 1960s, she began working with Theodore Rasmussen at the MNI, using the Wada technique to test lateralisation of function in the cerebral hemispheres – research showing how the hemispheric representation of language can vary as a function of handedness. Another part of the brain that Milner explored was the frontal cortex (a region that she particularly enjoyed testing in her lesioned patients). Her work would highlight the behavioural inflexibility and poor planning of individuals with frontal lobe damage and identify the dorsolateral region as important in the remembrance of time events (temporal memory). More recently, and well into her eighties and nineties, Milner has adopted the use of functional magnetic resonance imaging (fMRI) and PET scanning to examine the brain regions involved in spatial memory and language, including the neural basis of unilingual and bilingual speech.

As these words are being written, Milner is in her hundredth year and still working 3 days a week teaching and undertaking research. She is a professor at McGill University and remains at the MNI. Remarkably, her research career now spans over 75 years, and her only shortcoming (if newspaper interviews are to be believed) is she now sometimes forgets people's names. Brenda and Peter Milner divorced in 1970, although they remained the best of friends, living just a few blocks away from each other and speaking almost every day, until Peter's death at the age of 99 years in June 2018. Because she does not have a car, Milner walks to work most days. She says she has never bothered too much about money and enjoys nice food and good wine on occasion. Interestingly, Milner once remarked she likes working with men and has experienced little discrimination during her career. A recipient of numerous honours, too numerous to list here, special mention can nevertheless be made to the Companion of the Order of Canada, awarded in 2004, which recognises the highest service to Canada and humanity in general. Milner admits she is still curious and nosey – attributes that should inspire any aspiring psychologist keen in staking their claim to understand the brain and its cognitive functions.

Note

1 HM died on 2 December 2008, at the age of 82 – and only then was his real name (Henry Molaison) was publically revealed.

Milner's major writings

Milner, B., Corkin, S. and Teuber, H.-L. (1968) Further analysis of the hippocampal amnesic syndrome: 14 year follow up of H.M. *Neuropsychologia*, 6, 215–234.

Milner, B. (1982) Some cognitive effects of frontal-lobe lesions in man. *Philosophical Transactions of the Royal Society of London, Series B*, 298, 211–226.

Milner, B., Corsi, P. and Leonard, G. (1991) Frontal-lobe contribution to recency judgements. *Neuropsychologia*, 29, 601–618.

Milner, B. (1998) Brenda Milner. In: L.R. Squire (ed.), *The History of Neuroscience in Autobiography, Vol 2*. New York: Elsevier.

Milner, B., Squire, L.R. and Kandel, E.R. (1998) Cognitive neuroscience and the study of memory. *Neuron*, 20, 445–468.

Rasmussen, T. and Milner, B. (1977) The role of early left brain damage in determining lateralization of cerebral speech functions. *Annals of the New York Academy of Sciences*, 299, 355–369.

Scoville, W.B. and Milner, B. (1957) Loss of recent memory after bilateral hippocampal lesions. *Journal of Neurology, Neurosurgery and Psychiatry*, 20, 11–21.

Further reading

Banich, M.T. (2004) *Cognitive Neuroscience and Neuropsychology*. Boston, MA. Houghton Mifflin.

Corkin, S. (2013) *Permanent Present Tense: The Man with No Memory and What He Taught the World*. London: Allen Lane Books.

Kolb, B. and Whishaw, I.Q. (2015) *Fundamentals of Human Neuropsychology*. New York: Macmillan.

Rolls, G. (2005) *Classic Case Studies in Psychology*. Abingdon: Hodder Arnold.

Squire, L.R. (2009) The legacy of patient HM for neuroscience. *Neuron*, 61, 6–9.

Xia, C. (2006) Understanding the human brain: A lifetime of dedicated pursuit. Interview with Dr Brenda Milner. *McGill Journal of Medicine*, 9, 165–172.

Chapter 19

James Olds (1922–1976)

James Olds will always be associated, along with postgraduate student Peter Milner, for discovering the pleasure centres and reward systems of the brain – a breakthrough that would begin to transform our understanding of motivation, emotion and addiction.

James Olds was born in Chicago and grew up in Nyack, New York, and Washington, DC. It was an illustrious academic family. His grandfather, George D. Olds, had been professor of mathematics at the University of Rochester (1884–1891) and president of Amherst College (1923–1927). The two sons that followed him also had distinguished lives. George D. Olds II would found Continental Oil (later to be Conoco) and Leland Olds (father of James) became a distinguished economist and editor of the left-wing newspaper *Federated Press*, who had moved to Nyack in 1931 to head the New York State Power Commissions. He was later appointed by Franklin D. Roosevelt to chair the Federal Power Commission in 1939. Olds's maternal side of the family was no less impressive, with his mother having been the first female graduate of the Oklahoma School of Mines with an engineering degree. James Olds began his university education at the University of Wisconsin where the famous philosopher and educational reformer Alexander Miekeljohn was president, but he transferred after 1 year to St John's College in Annapolis, Maryland. It appears that Olds was not fully focused on his studies, however, since he took a year out to be a reporter for the International News Service – and following this was drafted into the army in 1940 to serve in World War II. Olds would spend most of his time in Tehran and Cairo with the Persian Gulf Command – a branch of the army that was assigned with the mission of expediting the shipping of war materials to the Soviet Union as part of the Land-Lease program. Olds would not only become fluent in Arabic but walk away unscathed from an air crash that claimed the life of several generals. He also managed to climb the highest mountain in Iran (Mt. Damavand), which is no mean feat at over 5000 m.[1]

After 3 years of service, Olds returned to the United States to complete his BA degree in philosophy magna cum laude from Amherst College in 1947. He then moved to Harvard University, joining the newly created Department of Psychology and Social Relations, which had been established by the sociologist Talcott Parsons. The department concentrated on teaching social, experimental and clinical psychology, along with sociology, and here Olds took his MA degree in 1951. He followed this with a PhD in 1952, under the supervision of comparative psychologist Richard Solomon, examining the behavioural aspects of motivation and emotion. During this time, Olds also made use of his journalist skills by taking on a part-time job as book editor of *Family, Socialization and Interaction Process* by Talcott Parsons. In fact, his contributions were so substantial that Parsons made Olds a co-author. It was not a typical apprenticeship for someone who would later make his reputation as a famous neuroscientist!

At Harvard, Olds had been influenced by Donald Hebb's book *The Organization of Behavior* – a work emphasising the importance of linking psychological theory to the workings of the brain and nervous system. Keen to adopt this biological approach, Olds won a fellowship to work with Hebb at McGill University. Although he had developed few, if any, physiological techniques to examine the brain, Olds nonetheless went to Montreal a "convinced neuroscientist". It may also have helped Olds's cause that his background was in the psychology of motivation and learning – subjects very much in vogue at the time. Hebb was not renowned for being an overbearing supervisor and left Olds to his own devices, giving him free reign to use a laboratory and the help of an undergraduate student. But perhaps most important of all, Hebb placed Olds under the guidance of Peter Milner who was then working on his own doctoral thesis. An experienced laboratory technician who built much of his own equipment, Milner would introduce Olds to a number of new techniques including stereotaxic surgery, electrode construction and the administration of electrical stimulation.

By now, Olds had developed an interest in the reticular activating system (RAS) – a diverse group of fibres extending from the brain stem to the thalamus, which then projected to widespread areas of the cerebral cortex. Although this system was known to be linked with sleep-wake cycles, there was some evidence it could affect learning by enhancing cortical arousal. In fact, Milner had already conducted pilot studies to see if electrical stimulation of the brain, including the reticular formation, could facilitate maze learning, and obtained disappointing results. But Olds was undeterred, and after learning the basics of stereotaxic surgery from Milner, he performed his first operation on a rat with a simple electrode made from standard insulated wire, aimed at the upper portion of the reticular formation. According to one biography, the planned experiment was to take place on a Monday, but on Sunday Olds decided to check the equipment was working adequately – and to do this, he placed the rat into an open arena and began passing short bursts of electrical activity into its brain with a hand-held button. Olds soon realised this was having an unexpected effect: the rat was always returning to the place in the open field where the last stimulus had been administered and seemingly sitting there patiently for more stimulation.

At first, Olds assumed the stimulation was producing a sense of curiosity in the rat. But a second day of testing showed otherwise. After training the rat to regularly visit a particular corner for stimulation, Olds then began trying to coax the animal into another corner by administering a current whenever the rat moved in that direction. Within a few minutes, the rat had moved to a new corner where it sat. As Olds later wrote in *Scientific American*, "After this, the animal could be directed to almost any spot in the box at will of the experimenter". The obvious interpretation was that the rat found the stimulation of its brain pleasurable, making it intent on finding the correct place to receive more of the same.

To confirm the rewarding nature of the procedure, Olds ran another test, by placing the animal on a T-shaped platform and giving brain stimulation if it only turned into the right arm. The rat soon learned to make the right turn on every trial. Then, Olds made the rewarding stimulation conditional on going left and found that, while the rat persisted for some time with its right preference, it eventually took to turning left. Olds also wondered if the electrical stimulation produced an effect stronger than hunger – and to examine this, he deprived the animal of food for 24 hours. When the rat was placed back onto the T-platform, he baited both arms with food – but this time administered the stimulation at a location halfway down the right arm. At this point, the rat showed no interest in the food and remained at the site where the stimulation was being administered.

Although these results appeared to show the rewarding nature of brain stimulation, Milner remained sceptical of this explanation. After all, he had undertaken his own studies prior to Olds arrival in the laboratory with negative results. According to Milner's autobiography (see Squire 2014), he then remarked to Olds that he would only be convinced of the rewarding effects of the stimulation if the animal could be made to work for it in a Skinner box. It was a comment that set in motion a paradigm that was to revolutionise the neuroscientific study of reward mechanisms in the brain. Rather than controlling the situation themselves, Olds and Milner now devised a method that allowed the rat to "voluntarily" stimulate its own brain by pressing a large lever. Constructed by Milner (McGill was according to Milner the only psychology department in the land without a Skinner box), the rat received its stimulation from a lead passing through the chamber's ceiling. As Olds later recounted, the first time they tested an animal in this box, it ended any doubts that electrical stimulation was rewarding. The rat made responses regularly, every second or so, and when the experimenters turned off the current after 30 minutes, it continued pressing the lever seventy times before falling asleep. Olds and Milner quickly realised they had invented a powerful technique, which for the first time, allowed the reward systems of the brain to be mapped out in precise detail. Prior to Olds and Milner, the pleasure systems of the brain were just a hypothetical idea. Now, they had opened up a new door into understanding the brain, which would soon result in many thousands of papers being published on the neuroanatomy and pharmacology of reward.

Even so, Milner was still unable to replicate the effects that Olds had achieved with his sole rat, despite using the same stereotaxic coordinates to place the electrodes in the brain. He then began to have doubts about the exact placement of Olds's electrode. Although Olds was not yet prepared to sacrifice his star performing rat to perform the necessary histological analysis (it was still providing a rich amount of positive data), Milner was able to take the animal to an X-ray laboratory, when it became immediately clear the electrode had not gone to the site where it was supposed to have been. In fact, the tip of the electrode was in the septal area – a region of the brain closely connected with the hypothalamus. Thus, Olds in his first attempt at stereotaxic surgery had missed the RAS by about 4 mm – a considerable margin for the rat brain.[2] It was a highly fortuitous mistake. If Olds had accurately targeted the RAS, then his stimulation would not have been rewarding at all.

Old and Milner now began to map the brain areas that gave rise to self-stimulation, with the first results of their endeavour being published in 1954. They showed that electrodes placed in the septum, cingulate cortex and mamillothalamic tract all caused the rat to self-stimulate at high rates. This work, however, would mark the end of the collaboration. Milner still had his doctoral thesis to complete, and in 1955, Olds (along with his wife Marianne who was also now actively undertaking self-stimulation research with him) moved to the University of California in Los Angeles. The next 10 years would be a highly productive period for Olds who would now examine nearly every region of the brain for self-stimulation. This work showed that electrodes placed in the motor and sensory regions of the brain typically produced low response rates that rarely strayed above chance. Moderate levels of responding (in the region of 200 bar presses per hour) were elicited from the hippocampus, cingulate gyrus and nucleus accumbens, and higher levels were still from the septum, amygdala and anterior hypothalamus (where 500 presses per hour were not uncommon). But the region that produced the greatest rates of self-stimulation was the *lateral hypothalamus* – which sometimes gave rise to bar pressing rates in excess of 1000 presses per hour. In fact, one rat made 2000 responses per day for 24 consecutive hours before exhausting itself (Olds 1958).

As Olds examined all the self-stimulation sites in the brain, he realised they formed part of a large pathway called the medial forebrain bundle. In fact, this is a collection of over fifty separate fibre tracts, many with bi-directional projections, linking the forebrain (e.g. the limbic system, striatum and frontal cortex) with midbrain (e.g. ventral tegmental area and upper brain stem). Most strikingly, these neural tracts appeared to converge in the lateral hypothalamic region, which lay in the middle of its route between forebrain and midbrain. This region also gave rise to the highest rates of stimulation, and not surprisingly, it was soon regarded as the most important pleasure centre in the brain. This idea has changed in recent years – not least because it is now recognised that many fibres pass through the lateral hypothalamic region without making contact. Moreover, increasing evidence has shown that dopaminergic pathways originating in the midbrain tegmentum – especially those projecting to the nucleus accumbens – are even more critical in producing reward and pleasure. Indeed, virtually all drugs known to produce significant dependence (i.e. addiction), including cocaine, heroin, alcohol and nicotine, have been shown to work, at least in part, on this system.

In 1967, Olds moved to the Californian Institute of Technology in Pasadena where he became one of the first researchers to study single-unit responses in the brain (reticular formation, hippocampus, hypothalamus, etc.), with recording microelectrodes in freely moving rats using conditioning experiments. Fascinated with computers and electronics, Olds's innovations not only attracted many more researchers to this field of research but his results also lent further support for the idea that learning is a distributed process in the brain.

Said to be an urbane man with bright sparkling eyes and quick agile gestures and movements, accompanied with a keen sense of humour, Olds received many honours for his work, including election into the National Academy of Sciences when aged only 45. He was also held in great affection by the many people who worked with him. Tragically, Olds died in a swimming accident in 1976, on a vacation at Laguna Beach in Orange County, California, at the age of 54. According to the man who coined the term "affective neuroscience", Jaak Pankseep, the discovery of reward centres in the brain was one of the greatest neuroscientific discoveries of the twentieth century – and this breakthrough would allow all kinds of pleasures and addiction, from drug dependence and sex, along with eating behaviour, gambling and even humour, to be better understood. It is intriguing to think that a mistake of just 4 mm in the placement of an electrode performed during stereotaxic surgery on a rat, undertaken by a man who at the time was more educated in social psychology than neuroscience, could have such a profound impact on how we understand the basic emotions that drive our fundamental human behaviour.

Notes

1 I am extremely grateful to professor Jim Olds who kindly provided me with some of this information regarding his father.

2 I can vouch for the difficulty in performing stereotaxic surgery on the rat brain. I learned the procedure as a PhD student – and it took me about a year before I could perform the task with a high degree of accuracy.

Olds's major writings

Olds, J. and Milner, P.M. (1954) Positive reinforcement produced by electrical stimulation of septal area and other regions of the rat brain. *Journal of Comparative and Physiological Psychology*, 47, 419–427.

Olds, J. (1956) Pleasure centres in the brain. *Scientific American*, 195, 105–106.

Olds, J. (1958) Self-stimulation of the brain. *Science*, 127, 315–324.

Olds, M.E. and Olds, J. (1963) Approach-avoidance analysis of the rat diencephalon. *Journal of Comparative Neurology*, 120, 259–295.

Olds, J. (1969) The central nervous system and the reinforcement of behaviour. *American Psychologist*, 24, 114–132.

Olds, J. (1977) *Drives and Reinforcements: Behavioral Studies of Hypothalamic Functions*. New York: Raven Press.

Further reading

Disterhoft, J.F. (2016) Olds, James (1922–1973). *Encyclopedia.com*. www.encyclopedia.com/psychology/encyclopedias-almanacs-transcripts-and-maps/olds-james-1922–1973.

Leibman, J.M. and Cooper, S.J. (1989) *The Neuropharmacological Basis of Reward*. Oxford: Clarendon Press.

Milner, P.M. (1989) The discovery of self-stimulation and other stories. *Neuroscience and Biobehavioral Reviews*, 13, 61–67.

Milner, P.M. (1991) Brain-stimulation reward: A review. *Canadian Journal of Psychology*, 45, 1–36.

Milner, P.M. (2014) Peter M. Milner. In: L.R. Squire (ed.), *The History of Neuroscience in Autobiography, Vol 7*. Washington, DC: Society for Neuroscience.

Thompson, R.F. (1999) James Olds 1922–1976. Biographical Memoirs of the National Academy of Sciences, 77, 1–19.

Wauquier, A. and Rolls, E.T. (1976) *Brain-Stimulation Reward*. Amsterdam: North-Holland Publishing Co.

Chapter 20

Julius Axelrod (1912–2004)

Axelrod was a pharmacologist who discovered that certain neurotransmitters such as noradrenaline and serotonin are inactivated through the reuptake into presynaptic nerve terminals – a discovery paving the way for the development of modern antidepressant drugs.

Julius Axelrod (known to his friends as Julie) was born on the lower east side of Manhattan to Jewish Polish immigrant parents who were both illiterate in English. His father somehow managed to support the family by working as a humble basket-maker. Axelrod attended Seward Park High School, and then unable to afford college fees, he did what many other bright children from poor families managed to do: attend the tuition-free City College of New York, where he graduated with a bachelor's degree in biology in 1933. Rejecting his parent's hope that he would train as a rabbi, Axelrod applied to several medical schools but was accepted by none – a rejection undoubtedly the result of quotas at the time that limited the number of placements for Jews in medical school. As Axelrod once remarked in an interview to a newspaper reporter in 1970, "I wasn't that good a student but if my name were Bigelow I probably would have gotten in". This was also the time of the Great Depression, the worst economic downturn in the industrial world, and work difficult to find. After a brief stint as a laboratory technician at the New York University (NYU) School of Medicine, Axelrod obtained a position at the New York City Department of Health's laboratory of Industrial Hygiene, which involved testing the vitamin content of various fortified food products. It was also around this time that Axelrod lost the sight in his left eye due to an industrial accident, which resulted in him wearing a patch for the rest of his life – an event that also caused him to be deferred from World War II draft. Now fully able to concentrate on his career, Axelrod took night classes at NYU and completed a master's degree in chemistry in 1941. His thesis examined the chemical breakdown of enzymes in thiamine (vitamin B1)-fed rats.

Axelrod's career had little future prospect, and he may well have continued in his routine humdrum testing of foodstuffs for the rest of his working life if it was not for an unexpected opportunity that came his way in 1946. One of the most widely prescribed pain-killers of the time was acetanilide – but in high dosage it could cause a serious blood disorder in which the haemoglobin became deficient in transporting oxygen. Axelrod was asked if he would like to examine this problem, which had funding, working under Bernard Brodie, who was professor of pharmacology at NYU and a highly eminent researcher on drug therapy and metabolism. Axelrod took up the post at the Goldwater Memorial Hospital as laboratory technician and soon discovered that the breakdown of acetanilide in the human body produces a substance called aniline, which in turn oxidised haemoglobin. It was this change that reduced haemoglobin's capacity to carry oxygen. Following this discovery, Axelrod developed a simple assay for measuring aniline in urine and plasma, and even took high doses of acetanilide itself so he could detect aniline in his own urine. It was an exhilarating experience for Axelrod to be involved in cutting edge research and he now began to develop what would be a lifelong enthusiasm for pharmacological sciences.

But there would be a more significant breakthrough when Brodie and Axelrod discovered another breakdown product of acetanilide, called "N-acetyl-p-aminophenol", later known as acetaminophen, which had potent analgesic effects without the deadly disadvantage of impairing the haemoglobin's function. This discovery was published in 1948 (Axelrod's first scientific paper), and it advocated the use of acetaminophen as a non-toxic pain-killer. In fact, acetaminophen was to become the most widely used analgesic in the world – going on sale in the United States in 1955 as Tylenol and a year later in Europe as Paracetamol. It still remains one of the bestselling pain-relievers over 70 years later. Unfortunately, neither Brodie nor Axelrod shared in any of the profits of this now famous drug – as it never occurred to either of them to attempt patenting their discovery.

In 1949, Brodie moved to the newly formed NIH, which was then located in Washington, DC (although it would move to

Bethesda), where he became head of chemical pharmacology – and was followed by Axelrod. Few could have guessed in these early days of the NIH that it would quickly become one of the most fertile research settings in the world – and Axelrod was right at the heart of it, working in the now famous Building 3, whose inhabitants would give rise to five Nobel laureates and a large number of biomedical researchers who got elected to the National Academy of Sciences. In this new environment, Axelrod began examining the metabolism of caffeine, along with a class of drugs known as the "sympathomimetic amines", which included amphetamine, ephedrine and methamphetamine. Axelrod found that these drugs were metabolised in the liver by a variety of pathways that included conjugation, deamination and hydroxylation – all of which involved enzymatic processes. Interested to know where these chemical reactions took place, Axelrod separated the subcellular fractions of liver by centrifugation and identified the intracellular membranes such as the endoplasmic reticulum as their site of action. This breakthrough allowed Axelrod to examine the reactions in more detail, and he soon identified an important role for a previously unknown group of enzymes that required the action of a substance called "nicotinamide adenine dinucleotide phosphate" and molecular oxygen. In fact, Axelrod had discovered a family of microsomal enzymes responsible for many aspects of drug metabolism, which later became known as the cytochrome P450 group – now recognised as the most important enzymes in phase I metabolism in mammals. It was a classic advance in pharmacology.

At this point, Axelrod was past 40 years old, and despite his seminal discoveries, he remained a lowly research technician without a PhD. To make matters worse, there were several disputes with Brodie over the recognition and priority of Axelrod's work. His friends urged him to go back to university to pursue a PhD, but the difficulty of supporting a wife and two young sons made this venture untenable. Fortunately, George Washington University came to his rescue when they agreed that Axelrod's already published papers could count towards his doctoral thesis, and since he already had a master's degree, he would only need to take out a year in order to take a few requisite courses and prepare for the necessary examinations. In 1954, he took a leave of absence from NIH and enrolled for his PhD. A year later, at 42, he received his doctorate after completing his thesis on the metabolism of phenylisopylamines – a class of drug known for their hallucinatory properties.

With a PhD in hand, Axelrod was recruited by Seymour Kety, who had become the first scientific director of the NIMH in 1949, to head the department of pharmacology that was part of the newly formed Laboratory of Clinical Sciences. The aim of the NIMH was to bring together leading scientists to investigate the biological basis of psychiatric disorders, and Axelrod began his task by examining the metabolism of lysergic acid diethylamide (LSD), a drug that was then being implicated in psychotic behaviour. In 1957, Axelrod turned his investigations to epinephrine (adrenalin) after learning that a potential metabolite, called "adrenochrome",[1] might produce schizophrenic-like hallucinations. The suspicion was that schizophrenia might result from the abnormal metabolism of epinephrine – and Axelrod attacked the problem by searching for the enzyme that might be responsible for the drug conversion. This would lead to a significant breakthrough when Axelrod discovered a novel enzyme called "catechol-O-methyltransferase" (COMT), which broke down catecholamines including epinephrine and norepinephrine (noradrenaline). This discovery revolutionised the understanding of catecholamine metabolism since it was then believed that the catechols were only broken down through the enzyme monoamine oxidase. Soon after Axelrod's discovery, COMT was found to be ubiquitous in the brain – a finding that directly led to the use drugs to inhibit this enzyme with efficacy in Parkinson's disease. In addition, the use of COMT as a biological marker for providing evidence of catecholamine synthesis and breakdown in depression and schizophrenia became common experimental strategies.

Axelrod's list of achievements were already impressive (the discovery of paracetamol, the P450 enzymes and COMT), but in 1958, he turned his attention to another problem that would win him a Nobel Prize. At the time, little was understood about the metabolism of epinephrine and norepinephrine[2] after they had been released into the synapse. One clue came from acetylcholine that was known to be broken down in the neuromuscular junction by acetylcholinesterase. At first, Axelrod thought that COMT may serve a similar role for epinephrine – but this proved not to be the case. Somewhat perplexed, Axelrod then began a series of studies where he labelled epinephrine with radioactive tritium [^{3}H] – a procedure that allowed the concentration of epinephrine to be measured with great accuracy in the tissues and blood. After injecting epinephrine [^{3}H] into experimental animals, Axelrod obtained an unexpected result. While most of the labelled neurotransmitter was rapidly metabolised by COMT, a substantial proportion (30–40%) remained unchanged. Furthermore, this concentrated in the organs richly innervated by the sympathetic nervous system where it remained unchanged for several hours. In short, Axelrod had discovered a new mechanism for the inactivation of neurotransmitters, which involved the recapture of the transmitter back into the nerve endings – a process he termed *reuptake*.

But this was not all: soon after his seminal discovery, Axelrod showed that the newly introduced tricyclic antidepressants, which were then just beginning to radically change the treatment of depression (the first tricyclic antidepressant Tofranil was first approved in the United States in 1959), were potent inhibitors of norepinephrine uptake – as was the psychostimulant cocaine. This advance not only provided a new window into helping understand the neurochemical basis of depression but it also led to the development of new antidepressant drugs. It also became clear that the reuptake process also took place in many other neurotransmitter systems including dopamine, glutamate, y-aminobutyric acid (GABA) and serotonin. The latter is especially important because serotonin has become the main target of today's generation of antidepressant drugs – the so-called selective serotonergic reuptake inhibitors as exemplified by fluoxetine (Prozac).

A final chapter in Axelrod's career concerned his work on melatonin synthesis and its regulation by the pineal gland. Prior to the late 1950s, the brain's pineal organ, which in humans is found in the third (or middle) ventricle, was a mysterious organ. That was, until the Yale chemist Aaron Lerner in 1958 showed that it produced the hormone melatonin (which he named). Axelrod was intrigued by melatonin because it contained an indole nucleus that is structurally similar to serotonin and also LSD. Working with Herbert Weissbach, Axelrod discovered that the pineal gland creates melatonin through an enzymatic modification of serotonin. He also realised that the release of melatonin followed a circadian pattern and that its rhythmicity persisted when the animals were placed in constant darkness, although abolished in constant light. This led Axelrod to develop an important theory that recognised the pineal gland as a transducer of information about the day/light cycle, which is turned into body's circadian rhythms, including the generation of the sleep-wake cycle.

In 1970, Axelrod, along with Ulf von Euler (who established that norepinephrine is a neurotransmitter) and Bernard Katz (who showed that neurotransmitters are released in discrete packages), were awarded the Nobel Prize for their "discoveries concerning the humoral transmitters in the nerve terminals and the mechanism for their storage, release and inactivation". Axelrod apparently first heard the news from his dentist – who mistakenly told him the award was for his "peace" work! Unlike many scientists, Axelrod's extraordinary discoveries had come from a small laboratory that normally only employed no more than three or four scientists. In fact, as one of his students (Solomon Snyder) once wrote, because of his many years as a technician, Axelrod felt uncomfortable if he was not himself carrying out an experiment. He also conveyed an infectious enthusiasm for making new discoveries to his students who inevitably came away from time in his laboratory having had fun and inspired from the experience. Described as a quiet, self-effacing mild-mannered man who rarely got angered, Axelrod retired in 1984 at the age of 72, although he continued to work in a small laboratory at NIMH well into his late eighties. He died of a heart attack in his home in 2004. Axelrod had been an inspiration for countless students and collaborators, and one, Michael J. Brownstein, would write that his genius "was not in doing experiments that required lots of technical finesse. It was doing work that anyone could have done – if they'd had the ideas. What separates the giants from the rest is the capacity to ask great questions".

Notes

1 Adrenochrome appears to have been a red herring. It is not a "normal" metabolite of epinephrine, nor is it currently believed to have any psychedelic properties.

2 These are known as adrenalin and noradrenaline in the United Kingdom.

Axelrod's major writings

Axelrod, J. (1957) O-methylation of epinephrine and other catechols in vitro and in vivo. *Science*, 162, 400–401.

Axelrod, J. and Hertting, G. (1961) The fate of tritiated noradrenaline at the sympathetic nerve endings. *Nature*, 192, 172–173.

Axelrod, J. (1988) An unexpected life in research. *Annual Review of Pharmacology*, 28, 1–23.

Axelrod, J. (2003) Journey of a late blooming biochemical neuroscientist. *Journal of Biological Chemistry*, 278, 1–13.

Brodie, B.B. and Axelrod, J. (1948) The fate of acetanilide in man. *Journal of Pharmacology and Experimental Therapeutics*, 94, 29–38.

Glowinski, J. and Axelrod, J. (1964) Inhibition of uptake of tritiated-noradrenaline in the intact rat brain by imipramine and structurally related compounds. *Nature*, 204, 1318–1319.

Moore, R.Y., Heller, A., Wurtman, R.J. and Axelrod, J. (1967) Visual pathway mediating pineal response to environmental light. *Science*, 155, 220–223.

Further reading

Craver, C.K.F. (2008) Axelrod, Julius. In: *Complete Dictionary of Scientific Biography*. New York: Charles Scribner's Sons, 2007 [ebook]. ISBN: 0-684-31559-9.

Iversen, L. (2006) Julius Axelrod. *Biographical Memoirs of the Fellows of the Royal Society*, 52, 1–13.

Kanigel, R. (1981) *Apprentice to Genius: The Making of a Scientific Dynasty*. Baltimore: John Hopkins University Press.

Pincock, S. (2005) Julius Axelrod. *The Lancet*, 365, 380.

Snyder, S.H. (2005) Julius Axelrod 1912–2004. *Biographical Memoirs of the National Academy of Sciences*, 87, 1–19.

Snyder, S.H. (2006) Turning off neurotransmitters. *Cell*, 125, 13–15.

Chapter 21

Rita Levi-Montalcini (1909–2012)

Rita Levi-Montalcini is famous for her discovery of nerve growth factor in the 1950s, which is essential for the growth, development and survival of neurons – work that led to a share of the Nobel Prize with Stanley Cohen in 1986.

Rita Levi (she later added her mother's maiden name Montalcini to her surname when beginning medical studies at the University of Turin in 1930) and twin sister Paola were born in Turin, Italy, to Jewish parents. Her mother was a painter – a pursuit also followed by Paola who became internationally known for her surreal and oneirically infused paintings. Rita's father was an electrical engineer, and by all accounts a caring but domineering man, who made all the household decisions which his wife accepted without challenge. This authoritarian disposition also extended to his daughters who were discouraged from pursuing a professional career and expected to follow their prescribed traditional female roles of getting married and bearing children. Nonetheless, from an early age, Rita vowed against marriage – a pledge she kept for the rest of her life. Instead, she was an avid reader of literature, never followed sports and found herself filled with a "profound sense of isolation" despite close ties with her mother. But Rita also had the support of several strong women in her family who were sources of inspiration, including two aunts who had doctoral degrees in literature and mathematics, respectively, along with a cousin Eugenia who was determined to have a university education. A formative moment came in 1929 when an employee of the Levi-Montalcini family, who had served as governess to the children, died from a poorly diagnosed gastric cancer. This was the moment when Rita, aged 20, decided to devote herself to studying medicine. Asking her father's permission to follow her dream, he was at first hesitant but then gave Rita his blessing. There were several obstacles to overcome – not least getting the qualifications that would enable her to enter medical school. But Rita, accompanied by her cousin, received the backing of their families who hired tutors, and they passed their entrance examinations with the highest scores after studying for just 6 months. They entered the University of Turin's medical school in 1930, being only two of seven women among the 300 or so first- and second-year students. Rita would, however, be in good company as two of her fellow students would include Salvador Luria and Renato Dulbecco, both of whom were to win the Nobel Prize in different fields of physiology.

At medical school, Levi-Montalcini came under the influence of leading histologist Giuseppe Levi who was well known for his fiery temperament. Rita later referred to him reverently as "the Master". Levi was also impatient and dismissive of students who did not learn quickly, although he was committed to those who proved worthy. Rita was keen to prove herself and joined his laboratory in the second year of her studies. At first, she examined patterns of neural growth in mice, which required her to laboriously count cells under the microscope. She was not initially keen on this type of work, although it was not long before she had become adept at using the silver impregnation method – a technique invented by Camillo Golgi in 1873, whereby silver is infused into samples of nerve tissue, allowing individual cells to stand out in entire clarity. Little could have Rita realised it at the time, how this technique would prove fundamental as one of the keys to her later success. Later in her studies, Rita was introduced to another important experimental procedure whereby nerve cells are removed from embryonic tissue and kept alive and growing in Petri dishes – a method allowing their development to be closely examined. Again, it would become another of Rita's specialities.

Levi-Montalcini completed her MD in 1936 and specialised in neurology and psychiatry – taking up an internship at Turin's Clinic for Nervous and Mental Diseases. She also began her own research observing the neural development of the chick embryo. But these were politically difficult times. In 1938, Czechoslovakia was occupied by Germany and the Italian dictator Mussolini cemented his anti-Semitic alliance with Hitler. This soon had serious implications for both Rita and Giuseppe Levi, when a law proclaiming that Jews were not members of the "Italian race" was passed in 1938. And soon after, it was followed by a royal decree excluding Jews from employment in universities. Forced from her job, Rita left for Belgium where she worked as a guest of a neurological institute in Brussels. But Brussels was no safer than Turin, and on the eve of the German invasion of Belgium in 1940, she returned to Italy. Realising it was no longer feasible to openly pursue research, Rita built a small laboratory in her bedroom. By this time, she had become intrigued by a paper written by Viktor Hamburger (who had fled from Germany in 1934) who was at Washington University in St. Louis, which described the effects of wing amputation or extirpation in the three-day-old embryonic chick – a procedure that caused up to 60% of the developing motor neurons to die. This was rather odd because the neurons were being formed by the spinal cord that was healthy. Despite only having simple tools for dissection, and she would eat the chicken eggs later for dinner, Rita replicated some of Hamburger's results using her silver staining technique.

Although agreeing with Hamburger's findings, Rita arrived at a different conclusion about the cause of cell death. Hamburger had proposed that the growing wing influenced the growth of the "distant" spinal neurons, by releasing a chemical that somehow travelled back along the primitive fibres, to make the neurons differentiate and mature. But Rita's observations suggested an alternative explanation: there was a chemical in the tissue of the wing bud that encouraged the growth of the neurons – and without it they soon died. On first sight, this appears to be a small difference – but it is not: Hamburger's theory suggests that neurons are programmed to develop in certain ways provided they receive the correct chemical signal. But Levi-Montalcini's theory is much more straightforward – holding that the wing secretes a factor that simply causes the neuron's growth towards that destination. Unable to publish her findings in an Italian scientific journal, Levi-Montalcini did manage to convey her views in a relatively unknown Belgium journal *Archives de Biologie* in 1943. A more extensive account would be published in the Italian journal *Pontifica Academia Scientarum* in 1944.

In 1941, Levi-Montalcini left Turin after heavy bombing of the city, and she set up her laboratory in a cottage in the countryside of Piedmont. But the German invasion of Italy in 1943 forced her to move again – this time to Florence where she remained in hiding until the Allies liberated the city in August 1944. When the war ended a year later, she returned to Turin and resumed her academic career. In 1947, she received an invitation from Hamburger to join him in St. Louis where he was professor and chairman of the zoology department. The purpose was to resolve their different interpretation of the wing ablation experiments. Although the stay was only planned for 1 year, the success of the work would soon lead Levi-Montalcini to become permanently employed at Washington University.

Soon after Rita had begun work, Hamburger received a paper from one of his former graduate students, Elmer Bueker, who had carried out his own limb transplantation experiments. Bueker had grafted small pieces of tumour material into chicken embryos and found that one type, known as mouse sarcoma 180, caused the developing spinal cord to infiltrate the tumour with nerves. Intrigued, Levi-Montalcini replicated the study and made another discovery by showing the nerves never made functional contact with the cells of the tumour. It was as if the tumour was secreting a diffusible growth-promoting factor that attracted the nerve fibres – and to examine this possibility, Levi-Montalcini undertook a clever experiment: she grafted the tumour material onto the thin sac (amniotic membrane) that surrounded the chicken embryo. Thus, the tumour was not in direct contact with developing tissues of the embryo but shared the same blood supply. Nonetheless, the tumour still produced dramatic effects, albeit now directionless, on nerve growth. It was strong evidence, if not the first definite proof, that the tumour was releasing a soluble nerve growth-promoting factor (NGF) that diffused in the blood and travelled into the embryo.

Although the fertilised chick embryo model had served her well, it was time-consuming and often capricious. Moreover, Levi-Montalcini knew she would have a better chance of identifying and isolating the growth factor if she could cultivate the growth of neural ganglia tissue in a dish (*in vitro*) rather than using the whole embryo. Indeed, it would greatly speed up her work, allowing her to see results in hours and not days. Hence, in 1952, Levi-Montalcini with two white tumour-bearing mice concealed in her handbag, nourished with some apple slices, travelled to Rio de Janeiro to learn how to culture neurons from one of her former students, Hertha Meyer. The next 4 months would in Rita's words be one of the most intense periods of her life. After learning how to culture primitive ganglia in a dish, Levi-Montalcini placed small pieces of mice sarcoma close to the tissue. She was thrilled to discover that this elicited, within 10 hours or so, a sun-like outgrowth of nerve fibres from the ganglia with their highest density facing the tumour.

In 1953, a young brilliant biochemist called Stanley Cohen joined Levi-Montalcini, and they set about trying to characterise the NGF produced by the mouse sarcoma. Their first step was to produce a cell-free (liquid) extract that replicated the growth-promoting effects on embryonic neurons. Following this, they attempted to see if the substance was a nucleic acid or a protein. In order to do this, Cohen and Levi-Montalcini performed experiments with snake venom (a rich source of phosphodiesterase)[1] to destroy any nucleic acid. To their surprise, this produced even more neural outgrowth when applied to their culture – indicating that the NGF was in the snake venom as well. This also led to the suspicion that NGF existed in a natural form in the mouse salivary gland (in fact, snake venom is highly modified saliva, and the mammalian salivary gland an analogue of the snake venom gland), a guess that was proven to correct. This was the crucial breakthrough to isolation: because of the high amounts of NGF in saliva, Cohen and Levi-Montalcini were able to purify the molecule, which proved to be a protein.

In 1960, this research culminated with the publication of three classic papers. The first, by Cohen, described how the NGF protein was isolated along with an antiserum. The second, by Levi-Montalcini and her assistant Barbara Booker, showed that NGF was essential for the differentiation, growth and health of a wide variety of mammalian ganglia, which included human tissue. But the third was arguably the most revealing and dramatic. Levi-Montalcini and Booker injected NGF antibodies (i.e. rendering NGF biologically inactive) into newborn mice, rats, rabbits and kittens and discovered that most of the sympathetic ganglion were destroyed during development. It proved beyond doubt that NGF was essential for the growth and functioning of the nervous system.

Since the discovery of NGF in 1956, over a hundred more neurotrophic factors have been discovered and thousands of papers published examining new aspects of growth regulation in a variety of cells including neurons. This work has also shown that the receptors for these growth factors lie on the surface of cells, and they activate a class of intracellular enzymes known as the tyrosine kinases. These are crucial in regulating cell growth, differentiation and survival – and their dysfunction has been linked with the development of certain types of cancer. NGF has a particularly important role to play in the brain where it has been implicated in neurodegenerative conditions such as multiple sclerosis and Alzheimer's disease.

In 1986, the Nobel Prize in physiology or medicine was awarded jointly to Rita Levi-Montalcini and Stanley Cohen for their discoveries of growth factors.[2] Levi-Montalcini was 77 years old and only the fourth woman to win this accolade. Other awards bestowed on Levi-Montalcini during her career included the Albert Lasker Award and the National Medal of Science – the highest scientific award given by the US government. Levi-Montalcini also has the distinction of becoming the first ever Nobel laureate to reach their hundredth birthday. After making further important discoveries in the early 1960s, Levi-Montalcini would spend increasing amounts of her time in Italy, especially after becoming director of a cell biology laboratory in Rome for the National Council of Scientific Research in 1969. Formally retiring from Washington University in 1977, where she had worked for three decades, Levi-Montalcini later founded an institute to study neurodegenerative diseases in Rome, in 2002. The Nobel Prize turned Rita into an Italian national icon and she took to her role by participating in television debates, becoming a senator in the Italian government and writing her autobiography *In Praise of Imperfection*. Keen to help women in science, especially from the third world, she established the Rita Levi-Montalcini Foundation in 1992. Described by one colleague as "ambitions, autocratic, generous, possessive, aristocratic, demanding, preserving, insightful and totally dedicated to her work", this tiny, seemingly frail Jewish lady had overcome numerous obstacles to become a doctor, scientist and Nobel laureate. An inspiration to women everywhere, Levi Montalcini died on 30 December 2012, aged 103. She had once remarked "if you stop working, you are dead".

Notes

1 A phosphodiesterase is an enzyme that breaks a phosphodiester bond. These bonds are particularly important in forming the backbone of the strands of DNA and RNA. Hence, a phosphodiesterase will break up the strands of nucleic acids.
2 The decision to leave Hamburger out of this award generated a great deal of controversy.

Levi-Montalcini's major writings

Levi-Montalcini, R. and Booker, B. (1960) Excessive growth of the sympathetic ganglia evoked by a protein isolated from mouse salivary glands. *Proceedings of the National Academy of Sciences*, 46, 373–384.

Levi-Montalcini, R. and Booker, B. (1960) Destruction of the sympathetic ganglia in mammals by an antiserum to a nerve-growth protein. *Proceedings of the National Academy of Sciences*, 46, 384–391.

Levi-Montalcini, R. (1982) Reflections on a scientific adventure. In: D. Richter (ed.), *Women Scientists: The Road to Liberation*. London: Macmillan.

Levi-Montalcini, R. (1986) The nerve growth factor: Thirty-five years later. In: T Frängsmyr (ed.), *Nobel Lectures, Physiology or Medicine 1981–1990*. Singapore: World Scientific Publishing Co. (1993).

Levi-Montalcini, R. (1988) *In Praise of Imperfection: My Life and Works*. New York: Basic Books.

Further reading

Abbott, A. (2009) Neuroscience: One hundred years of Rita. *Nature*, 458, 564–567.

Bradshaw, R.A. (2013) Rita Levi-Montalcini (1909–2012). *Nature*, 493, 306.

Cohen, S. (1960) Purification of a nerve-growth promoting protein from the mouse salivary gland and its neuro-cytotoxic antiserum. *Proceedings of the National Academy of Sciences*, 46, 301–311.

Dash, J. (1991) *The Triumph of Discovery: Women Who Have Won the Nobel Prize*. New York: Julian Messner.

Hitchcock, S.T. (2005) *Rita Levi-Montalcini: Nobel Prize Winner*. London: Chelsea House Publishers.

Iversen, L.L. (2013) Rita Levi-Montalcini: Neuroscientist par excellence. *Proceedings of the National Academy of Sciences*, 110, 4862–4863.

McGrayne, S.H. (2001) *Nobel Prize Women in Science: Their Lives, Struggles and Momentous Discoveries*. Washington, DC: Joseph Henry Press.

Yount, L. (2009) *Rita Levi-Montalcini: Discover of Nerve Growth Factor*. London: Chelsea House Publishers.

Chapter 22

Roger Wolcott Sperry (1913–1994)

Although Sperry made pioneering discoveries into the chemical basis of neural development, he is most famous for his research on split-brain patients – work showing that the two cerebral hemispheres have very different psychological abilities and personalities.

Roger Sperry was born in Hartford, Connecticut, to a father who worked in banking and a mother trained at business school, although he spent his early years on a farm where he developed a lifelong interest in nature. While it was not an ostensibly academic family, reading and scholarly achievement were valued in the home, and at the age of 10, Sperry was introduced to William James's book *Psychology* when his father bought it back from a public library. Unfortunately, Sperry's father was to die a year later, with his mother then securing employment at the local high school where she worked as an assistant to the principal. In his teenage years, Sperry excelled in athletics, achieve a state record in javelin and captain the school's basketball team. His interest in sports continued after winning a scholarship to Oberlin College, Ohio, at the age of 18, where his first ambition was to become an athletics coach. Despite this, Sperry would go on to get a BA in English literature. After graduating in 1935, Sperry remained at Oberlin, taking a master's degree in psychology under the guidance of Raymond Stetson, a follower of James and expert on speech therapy. But already Sperry was beginning to recognise the importance of understanding the neurobiological underpinnings of behaviour, as revealed in his first published scientific paper in 1939, which used a new instrument known as a kymograph to record muscular action. It was his Chicago move in 1938, however, to undertake a

PhD with the developmental neurobiologist Paul Weiss that led Sperry to greater things. This was a period dominated by behaviouristic thinking when it was widely believed the wiring of the nervous system largely came about through learning (as expressed by the dictum "anatomy arises from function"). But Sperry's doctoral research showed otherwise. Working with rats, he switched the flexor and extensor muscles of their hind limbs, or crossed the nerves to the same muscles. According to functional leaning theory, the animals should have adjusted to this change with time and experience. However, the rats showed no improvement in their awkward limb movements. These findings provided compelling evidence that the circuitry of the nervous system was predetermined and hard-wired for specific functions.

After completing his PhD in 1941, Sperry moved to Harvard to work under Karl Lashley, and soon after, he took up a post in the Yerkes Laboratories of Primate Biology also under Lashley. Although some of Sperry's work addressed issues concerning cerebral function, most notably Lashley's theory of equipotentiality (which Sperry rejected), his most important research continued with neuronal specificity – this time involving the rotation of eyes in amphibians such as newts and frogs. Previous work, dating back to the nineteenth century, had shown that cutting the newt's optic nerve resulted in nerve regeneration back to the brain's tectum when the eye was returned to its socket, thereby restoring full vision.[1] The question Sperry wanted to ask was: what would happen if the optic nerves were cut and the eyes rotated 180 degrees? Would the newt's vision return to normal or would it see the world as upside down? The results were clear: no matter the length of postoperative recovery, the animal saw its visual scene the wrong way round. Hence, when food was presented in the upper visual field, they looked down and vice versa. When Sperry traced the regenerating optic nerve, he found that the original pathway from eye to brain was maintained again – indicating that the growing retinal pathways were being guided to their original places of termination regardless of the eye's position. Again, it was strong evidence for the idea that the neural circuits of the brain are "hard-wired", with learning have little role in neural development. It also led Sperry to propose the chemoafinity theory that held that the optic nerve cells were tagged with a specific chemical, as was the tectum, which led the developing axon to its proper destination.

In 1946, Sperry returned to Chicago as assistant professor of anatomy, where he continued his nerve growth experiments, although he never obtained tenure and things got worse in 1949 when he was diagnosed with tuberculosis – most likely contracted at Yerkes after dissecting a monkey for nerve transplants. Thus, he was forced to spend 6 months in Saranac, New York, to recover. Sperry returned to Chicago as associate professor of psychology in 1952 and also held a position at the NIH in Bethesda. But perhaps because of his recuperation, Sperry was now taking a far greater interest in the neurological basis of the mind-brain problem. Indeed, his first account of the corpus callosum (albeit an abstract) was published in 1953 with doctoral student Ronald Myers. In 1954, Sperry was invited to give a talk at the Californian Institute of Technology (Caltech) in Pasadena. The lecture so impressed the faculty, Sperry was offered the first Hixon chair of psychobiology – a position he would hold until 1984.

At Caltech, Sperry focused his attention on the corpus callosum. This huge bundle of fibres that connect the two cerebral hemispheres had long baffled psychologists – not least, because it had been lesioned in the 1930s by doctors as a treatment for seizures without producing any obvious mental or behavioural impairment. This had even led Lashley, in 1951, to quip that the only known function of the corpus callosum was it stopped the two hemispheres from sagging! But in Pasadena, Sperry bought a greater sophistication to examining its function. Using cats as subjects, Sperry devised a means by which he could present visual information to each hemisphere. This is not simple as it first appears since each of the cat's hemispheres (as in a human) receives input from both the eyes. Thus, Sperry had to lesion the optic chiasm (where the optic nerves cross) and corpus callosum. Only then, could he arrange for each hemisphere to receive input from the same-sided eye. Now, by using simple eye patches, Sperry was able to selectively teach each half of the brain. His results showed that the learning curves for each hemisphere to learn visual discrimination tasks was identical – proving that two separate brains could exist side-by-side inside the cranium.

In the early 1960s, the Los Angeles neurosurgeon and visiting professor at Caltech, Joseph Bogen, provided Sperry with the opportunity to undertake split-brain research on human patients. The first subject was a World War II paratrooper known as WJ, who had been hit over the head with a rifle butt behind enemy lines

– an injury leading him to suffer from progressively worsening seizures (up to twenty a day). In 1962, Bogen, along with Philip Vogel, performed a callosalectomy on WJ, which reduced the fits. At this point, Sperry, along with graduate students Michael Gazzaniga, tested WJ in the laboratory. This would lay the foundation for research that was to continue for another 25 years. One difficulty was that WJ still retained an intact optic chiasma. Thus, visual information, even if presented to one eye, would reach both sides of the brain. Fortunately, Sperry and Gazzaniga devised an ingenious procedure whereby WJ was required to stare at a fixation point in the middle of a screen, which was followed by a brief visual stimulus (a tenth of a second or less) to the right or left visual field of the eye. This exposure was long enough for WJ to start processing the stimulus but not turn his eyes to it. Because of the anatomical nature of the optic nerve, this meant a stimulus presented to the left visual field passed to the right hemisphere and vice versa. In addition, a second task was utilised where WJ was blindfolded and asked to identify objects placed in his hands. In humans, the pathways conveying the sensations of touch, unlike the visual system, completely cross to the opposite side of the brain.

Testing soon revealed interesting differences between the two hemispheres. Most obvious was the superiority of the left for language. For example, when a written word such as "spoon" or "key" was presented to the left hemisphere, WJ was able to read it, vocalise it and understand its meaning. No recognition occurred, however, when the word was sent to the right. In fact, WJ would typically report he had only seen a flash of light, or sometimes nothing at all. The left hemisphere also had the ability to scribble down answers to simple questions through its control over the right hand (the motor cortex of the left hemisphere controls the right hand and vice versa) unlike the right brain/left hand. But, WJ's right hemisphere was not without its own skills. For example, Gazzaniga showed that when the left hand was allowed to search by touch alone a tray of objects for a specified item (e.g. a spoon) after having it pictorially presented to the right hemisphere, WJ always chose the correct object. And this recognition occurred when WJ was unaware of visual input being presented to him. Prior to this work, the right hemisphere had long been regarded as the minor hemisphere – now it was recognised as having its own unique form of intelligence.

Following WJ, many more split-brain patients would be tested. Almost without exception, the left hemisphere was found to be superior at understanding written language, and it excelled at solving analytical problems that required reason and logic. Despite this, the right hemisphere was not entirely without language since it could understand certain words and even read simple sentences. The right hemisphere was also better at solving spatial puzzles – and outperformed the left at recognising faces and pictures. Indeed, the right hemisphere (as demonstrated by the actions of the left hand) showed superior ability at drawing pictures, learning mazes and completing jigsaws.

The two hemispheres also had different temperaments, with the right reacting to events more emotionally. This was shown when a nude picture was presented to the right hemisphere – an event that caused one subject to blush and giggle, though unable to explain their reaction. In some instances, this even led to the two hemispheres acting in a contradictory manner. One example of this occurred when a patient was asked to read a book while holding it in their left hand. Although they found the book interesting (the reaction of the left hemisphere), they kept putting it down – presumably because the right hemisphere was poor at reading and saw little point in holding the book. There was even one patient who pulled his trousers down with his right hand, only to pull them up with his left! Paul S, was another interesting patient, with a high degree of language competence in both hemispheres. When asked what job he would like, Paul's left hemisphere said draftsman, while the right replied racing driver. His two hemispheres also held completely different opinions about President Richard Nixon.

These findings were truly groundbreaking and attracted attention from neuroscientists and the wider public alike. In short, they showed that each cerebral hemisphere, when disconnected, has its own unique linguistic, perceptual, cognitive, emotional and motor abilities. They even appear to have their own consciousness, personality and memories. Not surprisingly, these findings have captured the imagination of many, and the notion of right and left hemispheric differences now become widespread in western popular culture. Despite this, it is important to note that Sperry believed it was far too simple to classify individuals as being right- or left-hemispheric dominant. Moreover, he emphasised that the combined effect of bi-hemispheric activity amounted to more than the simple additive effects of the two separate hemispheres.

It is often overlooked that Sperry was also a highly original philosophical thinker, whose deeper reflections on his split-brain studies first took shape in 1965 when he published the first of a series of papers on consciousness, in which he advocated a new mentalist monist theory of mind. This broke away from reductionist views then prevalent in psychology. According to Sperry, the mind is an entity that "emerges" out of the chemistry and biophysics of nervous activity. Thus, while the mind is dependent on neural functioning, it is also something entirely different, which perhaps can be said to be created "above" the brain's biological processes. In order to illustrate this, Sperry made a parallel with quantum physics in which "interactions among subatomic particles result in emergent properties which in no way resemble the particles which they arose". On first sight, this appears like dualism that regards the mind as a separate entity from the physical brain. But Sperry insisted it is not. Rather, his theory is a monistic one (i.e. the brain and mind are composed of the same fundamental substance). Moreover, by adopting this approach, Sperry is also able to use the argument of downward causation as a means by which the mind can alter the functioning of the brain.

In 1981, Sperry was jointly awarded the Nobel Prize in physiology or medicine "for his discoveries concerning the functional specialisation of the cerebral hemispheres" (the other half of the prize went to Hubel and Wiesel). Sadly, by the time of his Nobel Prize, Sperry was suffering from a neural degenerative disease that affected his mobility and speech, although his intellectual function was spared. Consequently, his acceptance speech was read for him. Sperry continued at Caltech until his retirement in 1984. By this time, he had won numerous high awards and honours for his neurospecificity and split-brain achievements. Those who knew Sperry described him as a quiet, socially reticent person with a rich private life, which included numerous adventures such as camping on wild beaches, deep sea fishing and searching for giant ammonites and dinosaur bones. In fact, when the announcement of his Nobel Prize was made, he was on a beach in Baja, California, with his wife, stranded by a hurricane. Said by his students to have a wry sense of humour, who often kept a low profile by opting to stay in his office, often with feet on desk and "thinking", Sperry died in 1994 at the age of 80 in Pasadena where he had remained emeritus professor.

Note

1 This effect is generally only found in lower animals such as amphibians like frogs and newts. Mammals are unable to regenerate a severed nerve pathway

Sperry's major writings

Myers, R.E. and Sperry, R.W. (1953) Interocular transfer of a visual form discrimination habit in cats after section of the optic chiasma and corpus callosum. *Anatomical Record*, 115, 351–352.

Sperry, R.W. (1963) Chemoaffinity in the orderly growth of nerve fibre patterns and connections. *Proceedings of the National Academy of Sciences*, 50, 703–710.

Sperry, R.W. (1964) The great cerebral commissure. *Scientific American*, 210, 42–52.

Sperry, R.W. (1965) Mind, brain and humanist values. In: J.R. Platt (ed.), *New Views of the Nature of Man*. Chicago: Chicago University Press.

Sperry, R.W. (1977) Forebrain commissurotomy and conscious awareness. *The Journal of Medicine and Philosophy*, 2, 371–388.

Sperry, R.W. (1982) Some effects of disconnecting the cerebral hemispheres. *Science*, 217, 1223–1226.

Further reading

Bogen, J. (1999) Roger Wolcott Sperry (20 August 1913 – 17 April 1994). *Proceedings of the American Philosophical Society*, 143, 491–500.

Hamilton, C.R. (1998) Paths in the brain, actions of the mind: Special issue in honor of Roger W. Sperry. *Neuropsychologia*, 36(10), 953–954.

Ottoson, D. (1981) Sperry has given us a new dimension on views of the higher functions of the brain. *Lakartidningen*, 78, 3765–3773.

Todman, D. (2008) History of Neuroscience: Roger Sperry (1913–1994). IBRO History of Neuroscience. www.ibro.info/Pub/Pub_Main_Display.asp?LC_ID=3473.

Trevarthen, C. (1990) *Brain Circuits and Functions of the Mind: Essays in Honor of Roger W. Sperry*. Cambridge: Cambridge University Press.

Voneida, T.J. (1997) Roger Wolcott Sperry 1913–1994. *Biographical Memoirs of the National Academy of Sciences*, 71, 315–331.

Chapter 23

David Hunter Hubel (1926–2013) and Torsten Nils Wiesel (b. 1924)

Hubel and Wiesel will always be known for characterising the responses of single cells in the primary visual cortex, although they made many other groundbreaking discoveries concerning cortical architecture, along with highlighting the importance of experience in the visual system's development.

David Hubel was born in Windsor, Ontario, to American parents from Detroit. His father, who was of German ancestry, worked as a chemical engineer for the local Windsor Salt Company, and Hubel's mother was an independent Protestant with an interest in electronics. When Hubel was 3 years old, his parents moved to Montreal and settled into a middle-class, predominantly French, neighbourhood. From his early childhood, Hubel took an interest in science, carrying out chemical studies in the basement of his home and showing an aptitude for electronics and making radios. He also played the piano and claims his music teacher instilled in him a love of Bach that he would not have traded for any amount of success in science. In 1943, Hubel enrolled at McGill University, taking a BSc degree in mathematics and physics, although he chose not to follow these subjects after graduation. Instead, "almost on a whim" Hubel decided to study medicine. It proved to be a serious challenge since Hubel had never studied biology and would describe his initial experience as "like a blow to the jaw". Despite this, Hubel enjoyed biochemistry, and in his second year, he began to develop a strong interest in the brain. After earning his MD at the age of 25, Hubel took up a residency in neurology at the MNI, where he also worked in the laboratory of Herbert Jasper learning about EEG. There was also a requirement to give regular seminar talks, and Hubel was given the topic of vision – a subject he knew nothing about. Despite this, Hubel would come across papers by Keffer Hartline and Stephen Kuffler, who had used microelectrodes to explore how neurons in the optic nerve encode sensory information – work that impressed him greatly. In 1954, Hubel took up a residency at John Hopkins in Baltimore. The move from Canada to America now made Hubel subject to the army draft (as a doctor) requiring 2 years of service. Fortunately, Hubel was able to work this at the Walter Reed Army Institute of Research in Washington, DC, where he joined the neurophysiology group. Here, Hubel worked on several projects, although the most important would be when he was given the ambitious task of trying to record from single cortical cells in the visual cortex of anaesthetised cats while they were awake and asleep. Hubel soon realised this would require him to develop a new type of tungsten electrode that was strong enough to penetrate the dura of the brain – a task that took him several months to achieve. It was sometime around this juncture, in 1956, Hubel first met Torsten Wiesel as a visitor to the Walter Reed laboratory.

Torsten Wiesel was born in Uppsala, Sweden. His father was a psychiatrist and head of a mental institution on the outskirts of Stockholm – and the family lived in the enclosed grounds of the hospital, along with the inmates who looked after the garden. Later, when Wiesel was 9 years old, his father became director of another mental asylum, Beckomberga Hospital, which was the largest institution of its kind in Scandinavia with 1200 patients. Again, the family lived within the grounds, although this time it was less than idyllic, with Wiesel recalling times when he could hear agonised screams from the nearby building housing the most severely ill. Despite being sent to a private school, Wiesel claims he was a lazy, mischievous student, interested mainly in sports – that is, until he reached the age of 17 when he decided on becoming a doctor. From then on, Wiesel studied intently and entered the Karolinska Institute to take medicine in 1946. It was a 7-year course and Wiesel developed an interest in neurophysiology after coming under the influence of Carl Gustav Bernhard who was an expert on vision. After completing his degree, Wiesel worked in Bernhard's laboratory where he published his first paper on the effects of anaesthetics in treating seizures. But with the intent on pursuing neurophysiology, in 1955, Wiesel had the good fortune of being invited to work in the laboratory of Stephen Kuffler at John Hopkins. Here, Wiesel joined the research team examining the receptive field properties of retinal ganglion cells – a subject on which Kuffler was an authority.

Soon after their first meeting, Hubel and Wiesel were to cross paths again in 1958 when they participated in a symposium to present the results of their research. Hubel by then had managed to record from single cells in the cerebral cortex using cats that were unanaesthetised and unrestrained. This was some achievement since it was the first time this type of recording had been achieved. Wiesel was examining single cell responses from the retina. Although a brief encounter, it was clear the two men had much in common. In 1958, Hubel's stay at the Walter Reed Army Institute ended. Although he had secured a post at John Hopkins, there was going to be a 6-month delay before his laboratory would be ready. Hearing of this, Stephen Kuffler asked Hubel if he would like to join him and Wiesel in the interim. It was to be the beginning of one of the greatest collaborations in the history of neuroscience.

Hubel and Wiesel decided to pool their respective skills by mapping the receptive fields (i.e. the part of the world "out there" to which the cell is responsive) of individual neurons in the visual cortex of anaesthetised cats. The groundwork for this type of work had been prepared by Kuffler in the 1950s, who had shone small spots of light onto the retina while recording activity from individual ganglia cells that make up the optic nerve. He had found that the receptive field of each ganglion cell was round in shape and had centre and surround regions. One such ganglion neuron was called an "on-off" cell, because if a spot of light fell in the centre of the field, it would excite the cell, while the same spot caused inhibition of neural activity when shone into the surround. Conversely, the reverse was true for "off-on" cells. A large spot of light that covered both the centre and surround was found to produce little neural activity – with the excitation cancelling out the inhibition. But Hubel and Wiesel were being much more ambitious than Kuffler – they were recording from microscopic neurons in the cat's visual cortex and much higher up in the visual hierarchy.

At first, Hubel and Wiesel were unsuccessful in evoking activity in the visual cortex by stimulating the eye with spots of light – that is, until they made a fortuitous discovery, when they caused a brisk discharge in one of the neurons after repositioning a glass slide into projector. Suddenly, Hubel and Wiesel realised that the cell was reacting not to spots of light but to an edge or line. It was a pivotal moment, and the two men were so excited with this breakthrough that they examined this cortical cell for 9 hours before stopping. Needless to say, they would examine many more cells in the visual cortex with this new type of presentation. These results were published in Hubel and Wiesel's first paper (1959) that showed slits of light (effectively lines) were the most effective stimuli for activating neurons in visual cortex – in contrast to the spots of light Kuffler had found worked best for the ganglion cells. More specifically, the receptive fields of the cortical cells in the visual cortex were elongated, not round, although they still maintained the excitatory centre and inhibitory surrounds, which were characteristic of ganglion cells. Consequently, an edge had to fall within the excitatory area of the receptive field for the cell to fire maximally.

Soon after this work was finished, Hubel and Wiesel moved to Harvard (1959) where they would continue their research until the late 1970s. Their collaboration soon resulted in another classic paper, this time published in 1962, which Hubel and Wiesel described as their "magnum opus". The most important theme in this article was the distinction between *simple* and *complex* cells. Simple cells were essentially the same as those described in their first paper. But complex cells, while still responsive to edges, had greater latitude in their "preferred direction". That is, the edge tended to elicit a brisk response if it fell anywhere within the neuron's receptive field. Furthermore, complex cells were sensitive to movement – often responding with sustained firing to any motion over large regions of the receptive field, compared to a very narrow boundary separating the excitatory and inhibitory regions for simple cells that preferred stationary input. Later, in 1965, Hubel and Wiesel reported a third type of cell, *hypercomplex*, which responded best to a line of limited length that had to fall within the confines of the receptive field. There was also an anatomical location for the cells types. Simple cells were found exclusively in layer 4 of the visual cortex (this layer receives input from eye and lateral geniculate nucleus), whereas complex and hypercomplex cells were found in layers 1 to 3 and 5 to 6.

Hubel and Wiesel also identified the presence of "orientation columns". In brief, when they lowered a recording electrode perpendicularly into the visual cortex, they found that the receptive fields of the cells (whether simple or complex) all had the same preferred line orientations. Thus, if a simple cell was found to respond to vertical lines, then all the other cells in the column did the same. Hubel and Wiesel (1963) also showed that if the recording electrode is moved sideways from one column to the next, the preferred axis of the orientation of lines then rotates, with each 0.05 mm of producing a rotation of 10°. Thus, a transverse of 1 mm in length across the cortex will typically reveal a set of orientation columns capable of detecting edges over 180°. Not only this, Hubel and Wiesel realised that the majority of these cells were also reacting to binocular input – although most had a greater preference for input from one of the eyes. Hubel and Wiesel called this "ocular dominance".

Over the next two decades or so, Hubel and Wiesel would significantly build on these findings and make further discoveries about the early stages of visual processing. One of their most important was to show that the visual cortex consists of cube-like units, which they called 'hypercoloumns'. Each of these is roughly a square millimetre in size and contains the full arrangement of orientation columns from 0° to 180° (containing simple, complex and hypercomplex cells) along with an array of orthogonal cells encoding for right and left ocular dominance. Later, in 1979, Margeret Wong-Riley at the University of California, San Francisco, discovered that the visual cortex also stained intermittently for an enzyme called "cytochrome oxidase", which gave it a polka dot appearance. In fact, the cells in these "blobs" were found to process colour information, and they led to a revision of the hypercoloumn concept, with blobs forming part of a more complex anatomical unit now known as a *cortical module*. It is estimated that the human primary visual cortex contains in the region of 2500 cortical modules – each processing information from a small part of the visual world.

Another seminal contribution of Hubel and Wiesel was to demonstrate that the nature of visual processing is hierarchical. So, for example, they showed how the receptive fields of large numbers of ganglion cells in the optic nerve could input to form a single cortical cell with a preferred line orientation. In turn, multiple aggregations of simple cells could provide the neural input to create a complex cell. Thus, as visual information passes through this hierarchy, the complexity of the neural representation increases. Presumably, this hierarchical processing continues as visual information is analysed beyond the visual cortex. Indeed, there is evidence to support this theory, with neurons in the association areas of the temporal lobes known to respond to a wide variety of objects and others to faces. This also raises some other interesting wider questions – not least, do we have a cell at the top of the hierarchy for everything in our visual world – such as a "grandmother cell", etc.? The answer is probably no – but the issues are too complex to be addressed here.

Hubel and Wiesel also undertook important work on the development of the visual cortex – and in particular the age at which the structure of the visual cortex was modifiable through experience. Using both cats and monkeys that had one eye sealed from an early age, Hubel and Wiesel showed that columns in the visual cortex receiving input from the sighted eye, took over columns that were normally governed by the deprived eye. Moreover, once these animals had passed through a critical sensitive period, they became incapable of achieving binocular vision again once full sight was restored. These findings helped explain deficits in human childhood vision, which arise from amblyopia (also known as lazy eye) in which an eye fails to develop normal visual acuity if it is not being used properly during early life.

The collaboration between Hubel and Wiesel flourished for 25 years and is brilliantly summarised in their book *Brain and Visual Perception* (2005), which also includes many of their famous papers along with commentaries. There is little doubt that their work has transformed the field of visual neuroscience and was recognised in 1981 with the Nobel Prize "for their discoveries concerning information processing in the visual system". By then, the two men had stopped working together – a split that Hubel felt was largely due to the increasing pressure to undertake administration and teaching. Indeed, he wrote in 2005: "In the end we felt like two horses continuing to drag the same plough over the same terrain year after year, with constant guilt feelings that we were neglecting one or another committee or were getting behind in letters of recommendation". In 1983, Wiesel left Harvard to join Rockefeller University, New York, to head its section of neurobiology. Later, in 1991, he became the president of the university and acted in that post until his retirement in 1998. Hubel remained at Harvard for the rest of his life and continued to work on the visual system with a number of other researchers – most notably Margeret Livingstone on the processing of colour and stereopsis in the extrastriate cortex. In 1988, Hubel became president of the Society for Neuroscience where he used his prestige to extol the great benefits of animal research. He died from kidney failure at the age of 87.

Hubel and Wiesel's major writings

Hubel, D.H. and Wiesel, T.H. (1977) Ferrier lecture on functional architecture of macaque monkey visual cortex. *Proceedings of the Royal Society of London B*, 198, 1–56.

Hubel, D.H. (1981) Evolution of Ideas on the primary visual cortex, 1955–1978: A biased historical account. In: T Frängsmyr (ed.), *Nobel Lectures, Physiology or Medicine 1981–1990*. Singapore: World Scientific Publishing Co.

Hubel, D.H. (1982) Exploration of the primary visual cortex, 1955–78. *Nature*, 299, 515–524.

Hubel, D.H. (1995*) Eye, Brain and Vision*. New York: Scientific American Library.

Hubel, D.H. (1998) David H. Hubel. In: L.R Squire (ed.), *The History of Neuroscience in Autobiography. Volume 1*. New York: Academic Press.

Hubel, D.H. and Wiesel, T.N. (2005) *Brain and Visual Perception*. Oxford: Oxford University Press.

Wiesel, T.N. (1981) The postnatal development of the visual cortex and the influence of the environment. In: T Frängsmyr (ed.), *Nobel Lectures, Physiology or Medicine 1981–1990*. Singapore: World Scientific Publishing Co.

Further reading

Alonso, J.-M. (2009) My recollections of Hubel and Wiesel and a brief review of functional circuitry in the visual system. *Journal of Physiology*, 587, 2783–2790.

Barlow, H.B. (1982) David Hubel and Torsten Wiesel: Their contributions towards understanding the primary visual cortex. *Trends in Neurosciences*, 5, 145–152.

Kandel, E.R. (2009) An introduction to the work of David Hubel and Torsten Wiesel. *Journal of Physiology*, 587, 2733–2741.

Shatz, C.J. (2013) David Hunter Hubel (1926–2013): Neuroscientist who helped to reveal how the brain processes visual information. *Nature*, 502, 625.

Wurtz, R.H. (2009) Recounting the impact of Hubel and Wiesel. *Journal of Physiology*, 587, 2817–2823.

Wurtz, R.H. (2016) David Hunter Hubel. 27 February 1926–22 September 2013. *Biographical Memoirs of the Fellows of the Royal Society*, 62, 233–246.

Chapter 24

Arvid Carlsson (1923–2018)

Carlsson is famous for discovering that dopamine acts as a neurotransmitter in the brain, recognising that it plays an important role in the pathophysiology of schizophrenia and Parkinson's disease, and establishing the pharmacological action of antipsychotic drugs.

Carlsson was born in Uppsala, Sweden, to parents who were both historians. His father, Gottfrid, specialised in Nordic politics of the late Middle Ages and was appointed professor of history at the University of Lund when Carlsson was 3 years old. Carlsson's mother Lizzie, also a doctorate, was interested in the legal status of Swedish women during the same historical period – although she had given up her career aspirations to raise a family. Despite this, she had kept a keen interest in her subject and during her seventies would publish two books and a number of articles on historical matters, which resulted in a honorary degree from the University of Uppsala. Carlson recounts that he grew up in a loving stable environment with his two brothers and sister, and all of whom would have a university education. Carlson chose to study medicine as did his younger brother – although he had no real motive to do this, other than having a vague idea of believing science might be more useful than arts. He began his medical studies in 1941, at the University of Lund, in the "deep south" of Sweden, a town that Carlsson thought was somewhat reminiscent of Oxford. At this time, World War II was raging in much of Europe, and Carlsson's study became interrupted for a couple of years because of military service. Returning to clinical training in 1944, Carlsson was also one of the first medics to examine the emaciated prisoners of Nazi concentration camps whom had managed to reach Sweden through the humanitarian efforts of Count Folke Bernadotte.[1] Right from the start of his medical studies, Carlsson had a hankering to do research, and in the same year, he obtained an unpaid position as an assistant in the laboratory of pharmacologist Gunnar Ahlgren. Here, Carlsson was given the task of examining the action of the convulsant pentylenetetrazol. It would lead to his first paper in 1946. Carlsson then studied the stimulant nikethamide (Coramine),[2] a drug used medically to counteract tranquilizer overdoses. He found that its lethal action followed a circadian rhythm – the first time a drug's action had been shown to be influenced by time of day. Another area of research that Carlsson entered into (around 1948) concerned the effects of calcium on bone metabolism. In fact, this work would form the basis of his MD thesis, which he finally completed in 1951.

Although Carlsson had published several papers on calcium metabolism, he did not find it an appealing option for further research, and so he sought a laboratory in the United States where he could gain more experience of chemical pharmacology. This led to an invitation from Bernard Brodie in 1954 to work at the National Health Institute in Bethesda, Maryland. This was a period when psychopharmacology was undergoing a revolution with the emergence of the first drugs to treat mental illness, which included the antipsychotic chlorpromazine that had just been approved for use in psychiatric patients. Carlsson's task was to examine another antipsychotic agent, reserpine, which had first been isolated in 1951 from the snakeroot plant (*Rauwolia serpentina*). This substance had been used in India for hundreds of years to treat insanity and shown to be effective in the treatment of schizophrenia, although it also led to serious depression in many of its users. Although these side effects soon led to the discontinuation of reserpine's use, it was still of great importance to understand its pharmacological action. Brodie believed that the drug depleted the brain of the neurotransmitter serotonin (5-HT) and Carlsson was employed to test this theory by using blood platelets that were known to absorb 5-HT. Indeed, Carlsson's research would demonstrate a clear-cut blockade on the storage of platelet 5-HT by very low concentrations of reserpine (Carlsson et al. 1957).

After spending 6 months in Brodie's laboratory, Carlsson returned to Lund having been appointed associate professor of pharmacology. Here, he made contact with the professor of histology, Nils-Åke Hillarp, who had just discovered there were organelles in the neuroendocrine cells of the adrenal medulla that stored the catecholamines adrenaline and noradrenaline (NA). This was in some respects an adrenergic version of the platelet model, and Carlsson was keen to see if reserpine also affected the storage of these neurochemicals. In fact, Carlsson found that injections of reserpine into experimental animals such as mice and rabbits caused a depletion in adrenaline and NA from the adrenal tissues. And Carlsson confirmed this when he stimulated the sympathetic nerves innervating the adrenal medulla and found that the normal release of NA no longer occurred. In effect, the neurotransmitter had gone from the axon endings – and this loss was soon extended to the CNS when Carlsson found reserpine depleted the brain's storage of NA. It was a provocative finding and it raised another question: what was more important for explaining reserpine's action – NA or 5-HT? To address this, Carlsson and Hillarp injected animals with reserpine, which led to immobility and marked sedation, and then assessed if these behavioural effects could be reversed by enhancing 5-HT or NA. Because peripheral injections of 5-HT and NA would not have crossed the blood-brain barrier, Carlsson and Hillarp administered their respective precursors instead – namely, 5-hydroxytryptophan and L-Dopa. The most striking effect was obtained with L-Dopa that completely reversed reserpine's sedative effects after just 10 minutes.

It might have been expected from this result that L-Dopa had somehow acted to restore normal levels of NA again. But this was not the case. Curiously, the brains of the experimental animals still showed marked depletion of NA, even though normal behaviour had been restored. Somewhat puzzled, Carlsson and Hillarp now turned their attention to dopamine – at the time, a recently discovered chemical in the brain whose main biological role was believed to be as an intermediary in the formation of NA. It was an inspired decision because Carlsson and Hillarp not only found that reserpine depleted the levels of dopamine in the brain but there also was an almost perfect correlation between the behavioural effects of reserpine and the loss of dopamine. It was also a result that tentatively indicated that dopamine might have a neurotransmitter role in the brain independent of NA. To test this possibility, Carlsson began to measure levels of dopamine in the brain under normal conditions. Although he was not the first to do this (dopamine had been identified in the brain a year previously), Carlsson showed that the concentrations of dopamine were comparable, if not even higher, than those of NA. Put another way, the concentration of dopamine indicated that it acted as a neurochemical in its own right (Carlsson et al. 1958). Shortly after this discovery, Carlsson's laboratory showed that dopamine's distribution in the brain was somewhat unusual – with most of it found in the striatum. This was an interesting finding because reserpine was known to produce Parkinson-like effects in high dosage. It did not take long for Carlsson to claim that a deficiency of dopamine in the striatum might be a cause of Parkinson's disease – a far-reaching claim that turned out to be correct (Carlsson 1959). This finding would lead directly to the treatment of Parkinson's disease with L-Dopa in the mid-1960s, which at the time was arguably neuroscience's greatest success story (and perhaps still is!). Today, L-Dopa remains the number one drug treatment for this disorder, and Carlsson would perform the first clinical study of L-Dopa for Parkinson's in Sweden in 1970.

During the early part of the 1960s, Carlsson turned his attention towards understanding the pharmacological action of the antipsychotic drugs chlorpromazine and haloperidol, which were then being used extensively to treat schizophrenia. Both these drugs produced side effects similar to reserpine, including sedation and immobility – yet, neither depleted the brain of dopamine. Working with Margit Lindvist, who had first introduced Carlsson to laboratory techniques as a young medical student in 1944, Carlsson discovered that the administration of either chlorpromazine or haloperidol increased the levels of dopamine breakdown product 3-methoxytyramine in the striatum while leaving the dopamine levels unchanged. This was an odd finding – although one that could be explained using a leap of imagination. In short, Carlsson and Lindvist proposed that instead of reducing the stores of dopamine (as reserpine does), chlorpromazine and haloperidol exerted their effects by blocking postsynaptic dopamine receptors. In turn, this blockade led to a feedback signal instructing the presynaptic neurons to release more neurotransmitter. This was a radical idea not only because it advocated receptor blockade but also because it was the first time a receptor-mediated feedback control had been suggested to act in the nervous system. Again, Carlsson would be proven correct over the subsequent years by numerous workers using a variety of different techniques. These observations also formed the basis for the "dopamine hypothesis of schizophrenia", which saw the illness as resulting from abnormal activity of dopamine systems in the brain. Today, despite its limitations, this still remains the most viable theory of schizophrenia.

Another area of investigation to take place in Carlsson's laboratory during the 1960s was the development of histological techniques that identified the presence of catecholamines in the brain. At the beginning of the decade, the only effective way of visualising these chemicals was to use tissue taken from the adrenal glands. However, in 1961, Hillarp, along with a young doctor called Bengt Falck, discovered that it was possible to visualise catecholamines in the brain by exposing freeze-dried tissue to dry formaldehyde gas – causing a reaction that converted dopamine and NA into iso-quinoline molecules. These then emitted a yellow-green fluorescence under a microscope. This was the first method to enable sensitive visualisation of a neurotransmitter within nerve cells, and it dominated the field of catecholamine research for more than two decades. In fact, it was only replaced in the early 1980s by more versatile immunohistochemical methods that used antibodies to tag neurochemicals. The fluorescent technique soon allowed Carlsson and his colleagues to show that catecholamines were stored in the synaptic vesicles of nerve terminals – a finding confirmed by Hokfelt in 1968 with the use of the electron microscope. The visualisation of catecholamines not only helped prove the existence of chemical neurotransmission in the CNS but also led to the realisation that these neurochemicals were confined to distinct pathways. For example, a lesion of the substantia nigra led to the disappearance of green-fluorescent nerve terminals in the striatum, with a concomitant depletion of dopamine (as measured biochemically). Thus, this demonstrated the existence of a nigrostriatal dopamine pathway. Later modifications, notably the glyoxylic acid method, by Lindvall and Björklund in the 1970s greatly improved the sensitivity and precision of this technique and helped show that all the forebrain's NA, 5-HT and dopamine actually derived from cell groups in the midbrain.

Two other discoveries by Carlsson must also be briefly mentioned. First, in the late 1960s, Carlsson showed that tricyclic antidepressants not only blocked the reuptake of NA (as had been shown by Axelrod) but also of 5-HT as well. This would lead him to develop the first 5-HT uptake inhibitor called Zimelidine, which became the forerunner of many more selective 5-HT uptake blockers such as fluoxetine (Prozac) developed by the drug company Eli Lilly. Second, in the early 1970s, pharmacological research began to point to the possibility of receptors located on the presynaptic endings of catecholamine neurons that acted to regulate the release of their own neurotransmitters. Carlsson supported this theory by showing that the synthesis of dopamine was inhibited by certain dopaminergic receptor agonists and stimulated by antagonists. In 1975, Carlsson used the term *autoreceptors* to refer to this specialised presynaptic receptor – a word that has now become widely adopted. It is also now recognised that changes in autoreceptor function play a particularly significant role in the pharmacological action of many antidepressant drugs.

In 1994, Carlsson was awarded the Japan Award for his discovery of dopamine as a neurotransmitter and its clarification in mental illness and motor disease – an award that came with a huge prize of ¥50 million (nearly US$460,000). Six years later, in 2000, Carlsson was the recipient of the Nobel Prize (along with Eric Kandel and Paul Greengard) for his work concerning signal transduction in the nervous system. Carlsson is undoubtedly one of the greatest neuropharmacologists of the twentieth century, whose list of major discoveries has led to a paradigm shift in our understanding of the brain. These include: the discovery of dopamine in the brain, the first biochemical and pathophysiological understanding of Parkinson's disease leading to its modern treatment, establishing how antipsychotic drugs work so pivotal to the dopaminergic theory of schizophrenia, the recognition of autoreceptors and the development of the first selective 5-HT uptake blocker – and this list is far from exhaustive. Now in his nineties, he runs his drug company Carlsson Research and remains committed to research. Carlsson is also well known for his opposition to the fluoridation of drinking water and successfully lobbied the Swedish parliament to make this practice illegal. He is married with five children, one of whom (Maria) is assistant professor of pharmacology at Lund where she works with her father.

Sadly, Carlsson died in June 2018 at the age of 95.

Notes

1 Count Folke Bernadotte was a Swedish diplomat. During World War II, he negotiated the release of about 30,000 prisoners from German concentration camps, including 450 Danish Jews from the Theresienstadt camp in the city of Terezín, located in Czechoslovakia.

2 Coramine may have been one of the drugs used by suspected British serial killer and doctor John Bodkin Adams during the 1950s. It was

also used by Hitler's personal physician Theodor Morell, who injected it into the German leader whenever he became unduly sedated with barbiturates.

Carlsson's major writings

Carlsson, A., Shore, P.A. and Brodie, B.B. (1957) Release of serotonin from blood platelets by reserpine *in vitro*. *Journal of Pharmacology and Experimental Therapeutics*, 120, 334–339.

Carlsson, A., Falck, B. and Hillarp, N.-Å. (1962) Cellular localization of brain monoamines. *Acta Physiologica Scandinavica*, 196, 1–27.

Carlsson, A. (1987) Perspectives on the discovery of central monoaminergic neurotransmission. *Annual Review of Neuroscience*, 10, 19–40.

Carlsson, A. (1993) Thirty years of dopamine research. *Advances in Neurology*, 60, 1–10.

Carlsson, A. and Wong, D.T. (1997) A note on the discovery of selective reuptake inhibitors. *Life Sciences*, 61, 1203.

Carlsson, A. (1998) Arvid Carlsson. In: L.R. Squire (ed.), *The History of Neuroscience in Autobiography, Volume 2*. New York: The Society of Neuroscience.

Carlsson, A. (2003) A half-century of neurotransmitter research: Impact on neurology and psychiatry. In: H Jörnvall (ed.), *Nobel Lectures, Physiology or Medicine 1996–2003*. Singapore: World Scientific Publishing Co.

Carlsson, A. and Carlsson, M.L. (2006) A dopaminergic deficit theory of schizophrenia: The road to discovery. *Dialogues in Clinical Neuroscience*, 8, 137–142.

Further reading

Björklund, A. and Dunnett, S.B. (2007) 50 years of dopamine research. *Trends in Neurosciences*, 5, 185–187.

Healy, D. (2002) *The Creation of Psychopharmacology*. Cambridge, MA: Harvard University Press.

Lees, A.J., Tolosa, E. and Olanow, C.W. (2015) Four pioneers of L-dopa treatment: Arvid Carlsson, Oleh Hornykiewicz, George Cotzias, and Melvin Yahr. *Movement Disorders*, 30, 19–36.

Roe, D.L. (1997) From DOPA to Parkinson's disease: The early history of dopamine research. *Journal of the History of the Neurosciences*, 6, 291–301.

Roe, D.L. (1999) The discovery of dopamine's physiological importance. *Brain Research Bulletin*, 50, 375–376.

Chapter 25

Norman Geschwind (1926–1984)

Geschwind was a pioneer in behavioural neurology, who not only showed that many higher function deficits such as aphasia, apraxia and agnosia can be understood as disconnection syndromes in which there is damage to cortical neural connectivity but also demonstrated that the two human cerebral hemispheres are anatomically and functionally asymmetric.

Norman Geschwind was born and raised in Brooklyn, New York, and was the second son of Jewish parents who had emigrated from Polish Galicia (then part of the Austro-Hungarian Empire) some 20 years earlier. As a child, Geschwind attended a local yeshiva school, learning to speak Yiddish and some Hebrew, although these would be difficult times following the death of his father from pneumonia when he was just 4 years old. Later, Geschwind attended Boy's High School in Brooklyn, which he described as the most profound educational experience of his life. Here, he became fluent in French and Latin (he would ultimately be able to speak eight languages) while attracted to mathematics, which became his chosen subject when entering Harvard in 1942 on a Pulitzer Scholarship. Soon after beginning university, Geschwind took an interest in medicine – although his ambitions were thwarted when drafted into the army in 1944. Sent to Europe as an infantryman, Geschwind saw combat in Czechoslovakia and Germany. Then, in late 1945, Geschwind was dispatched to Japan where he became interested in the psychology of fighting soldiers, especially when they forgot their own safety and attacked at times when their lives were in great peril – an effect he attributed to the dissociation of cortical and limbic areas of the brain. After the war, Geschwind returned to Harvard, changed his major to psychology and completed it magna cum laude. He then enrolled into Harvard Medical School with the intent of becoming a psychiatrist – but came under the influence of Marcus Singer who introduced Geschwind to the neurological specialisations of aphasia and epilepsy. Graduating with a MD in 1951, and following an internship in Boston, Geschwind obtained a post as physician at London's National Hospital specialising in neurology. Now, fascinated with the study of neuromuscular diseases, Geschwind was greatly influenced by Sir Charles Symonds who stressed the importance of "localising" the lesion rather than "jump to a diagnosis". This reinforced, yet again, for Geschwind, the importance of understanding the brain in diagnosing neurological disorders.

On his return to the United States in 1955, Geschwind continued his neurological training at Boston City Hospital, and a year later (1956–1958), he became a research fellow at the Massachusetts Institute of Technology (MIT), spending some of his time undertaking neurophysiological studies of the giant squid neuron with Francis O. Schmitt. In 1958, Geschwind was appointed staff neurologist at the Boston Veterans Administration Hospital, where he worked with the distinguished neurologist and expert in language disorders Fred Quadfasel. A man who had been imprisoned for 2 months in Germany in 1935 for his opposition to Nazi rule, Quadfasel had also trained under Karl Bonhoeffer and Kurt Goldstein – both of whom had been pupils of the renowned Karl Wernicke. Quadfasel would greatly influence Geschwind by encouraging him to study the classical texts of neurology from the early nineteenth and early twentieth century, many of which were in German and French and which supported localisation theory – a position that Geschwind would also adopt as his own.

In Boston, Geschwind had access to a large number of patients with varying degrees of language impairment (aphasia) and this was the start of one of his lifelong scientific interests. In 1965, Geschwind helped establish the Boston University Aphasia Research Center – one of the first interdisciplinary institutes dedicated to aphasia research in the world. The Center brought together not only neurologists and psychologists but also those with other skills such as linguists and speech pathologists. Although Geschwind took up the chair of neurology at Boston University in 1966, he continued to be closely associated with the Aphasia Research Center. Later, in 1969, Geschwind was appointed professor of neurology at Harvard Medical School and would occupy this position for the rest of his life.

During the early 1960s, there were two schools of thought regarding the effects of brain lesions on behaviour. The first were the anti-localizationists who believed that higher mental functions including cognition, perception and language, cannot be localised to discrete brain regions. Thus, the adherents of this view held that the brain is holistic in its function, which was supported by a number of eminent brain scientists including Hughlings Jackson, Kurt Goldstein and, most famously, Karl Lashley. It had also been bolstered by the work of Andrew Akelaitis in the 1940s, who found few deleterious behavioural effects in patients with corpus callosum lesions for the treatment of epilepsy. Consequently, the anti-localizationism dominated neurological thinking in the early 1960s, and this influence extended to language disorders in which Geschwind was interested. However, when Geschwind joined the Veterans Hospital in Boston, there was change in the air. By the late 1950s, Sperry had shown that there were notable deficits in behaviour after severing the callosum in animals, and during the early 1960s, Gazzaniga would report the first results with split-brain humans, which showed dramatic differences in function between the hemispheres.

Around this time, Quadfasel gave Geschwind an original reprint of a paper published in 1892 by the French neurologist Jules Déjérine, reporting the case of a 68-year-old patient (Monsieur C), a highly educated businessman, who had suffered a stroke, which resulted in him being unable to read (alexia), although he could still write. More commonly known as *pure alexia*, it would lead Geschwind to develop a new concept about how to understand the effects of brain damage – which emphasised neural "disconnection". Déjérine's patient also exhibited some odd patterns of behaviour. For example, despite being unable to read, he could copy out words and understand their meaning if each letter was spoken out. Monsieur C was also blind in his right visual field, indicating that he had suffered damage to the left visual cortex – a prognosis confirmed at his autopsy. But even more important to explaining the reading deficit, for Geschwind, was the discovery of damage to the posterior portion of the corpus callosum. This implied that while Monsieur C could see the words and letters normally with his right visual cortex, the lesion of the corpus callosum stopped this information from accessing the language-dominant left hemisphere where reading comprehension took place. The ability to copy out written information could then be explained by intact connections between the right visual cortex and the motor areas that controlled hand movement. The implication, therefore, was that copying out letters involved an alternative route through the brain, which did not involve the left-sided reading areas. If this explanation was correct, then it not only supported an important role for the corpus callosum (supporting the work of Sperry) but also showed that disconnection of pathways between major brain areas was an important cause of neurological impairment – a position that also supported localisation of function.

Around the same time, one of Geschwind's colleagues, Edith Kaplin, came across a patient, (a 41-year-old policeman known as PK) who had undergone surgery to remove a tumour from his left hemisphere. Unfortunately, PK had also suffered a postoperative stroke. Although able to gesture and write with his right hand, PK could not write or type with the left. In an attempt to understand this more fully, Geschwind came across a long forgotten work by the German neurologist Hugo Liepmann, who had described a similar syndrome in 1905 in a patient with corpus callosum damage. From his study of the patient, Liepmann had theorised that the control of hand movements involved in writing must depend to a large extent on the transfer of information from the left hemisphere to the right cerebral motor areas, via the anterior corpus callosum. After reading this paper, Geschwind realised that PK's deficit was caused by a similar type of disconnection. Geschwind presented the case to the Boston Society of Neurology and Psychiatry in 1961 and predicted that PK most likely had sustained damage to the anterior four-fifths of the corpus callosum. Months later, a post-mortem confirmed Geschwind's predictions.

All of this evidence revealed to Geschwind that the brain should be regarded as a network, which conveyed information to and from important decision centres in either hemisphere. Consequently, Geschwind now turned to reassessing all the older literature, animal research and his own clinical experience in light of this new realisation. The results would appear in 1965 when Geschwind published a two-part paper in the journal *Brain* entitled "Disconnexion syndromes in animal and man". Its main aim was to outline a connectionist model of brain function, exploring its implications not only in the field of aphasia but also for many other neurological conditions including apraxia and visual agnosia. While the idea that brain dysfunction could be better understood as a result of disconnection of pathways, rather than focal cortical damage, was not new (although it had been largely forgotten), the clarity of Geschwind's exposition and force of argument persuaded many of its importance. It also dealt a serious blow to the holistic model of brain function. Instead, a number of neurological disorders could be better explained in terms of damaged pathways, many involving the corpus callosum, linking specialised cortical regions.

According to Geschwind, a *disconnection syndrome* occurs when a lesion of the myelinated white matter occurs to the commissural fibres, or to the association pathways within a hemisphere. In formulating this theory, Geschwind strongly emphasised the work of German Paul Flechsig who had established certain laws of cortical connectivity in 1901. Flechsig had shown that the primary sensory areas of the cortex (visual auditory, etc.) only make connections with their own respective adjacent unimodal association areas. But in turn, these areas project to higher order polymodal association areas ("the association areas of the association areas") that have a rich array of connections with many other regions of the cerebral cortex (and limbic system). This principle also applied to the linkages between the two hemispheres, since the fibres that pass through the corpus callosum derive from areas of secondary association cortex. Geschwind further stressed that the human brain is unique because of its rich intra-connectivity between association areas (a feature that does not occur in other animals whose cortex is more richly connected with the limbic system). Thus, for Geschwind, knowledge of this wiring is profoundly important for a functional understanding of the human brain and being able to accurately diagnose clinical syndromes arising from its damage.

Geschwind gave many examples of where this type of reasoning could be applied. Perhaps his most famous was in the case of *conduction aphasia*. This form of language impairment had first been predicted by Wernicke in 1874, who recognised that a pathway for language must exist, which connects a temporal lobe area for comprehension (now called "Wernicke's area") with Broca's area in the frontal lobe, known to be involved in speech production. Wernicke predicted that patients with conduction aphasia would have fluent speech but be liable to make word errors and grammatically incorrect sentences, although able to recognise their mistakes. Another deficit would be an inability to encode phonological information for production, as occurs when the patient is asked to repeat sentences. Geschwind's rediscovery of disconnection syndromes helped bring this aphasic syndrome back into clinical neurology – with Wernicke's predictions being largely confirmed. In addition, Geschwind described other disconnection syndromes involving Wernicke's area including those leading to tactile aphasia, pure word deafness, pure alexia and agnosia.

By the mid-1960s, it was clear from split-brain research that the two cerebral hemispheres had different functions. Geschwind also added to this corpus of knowledge. For example, in a 1968 study with Walter Levitsky, he examined 100 human brains at autopsy. This revealed that two-thirds had a larger left-sided planum temporale – a region that includes part of Wernicke's area (the mean length of the left planum was 3.6 cm and of the right was 2.7 cm). Later, in the 1970s, with Albert Galaburda, Geschwind showed that neural anomalies of the planum temporale were associated with dyslexia. But perhaps most intriguing, collaborating with Peter Behan in London, Geschwind found in a random sample of 500 left-handers and 900 right-handers that dyslexia was much more likely in left-handers – an effect he linked to a higher amount of immune disease. And, more provocatively, he also argued this overall effect could also be explained by the release of higher levels of testosterone during the last trimester of foetal development – a theory that continues to generate much research.

Sadly, Geschwind's life was tragically cut short, when at the peak of his scientific productivity he died from a heart attack in 1984. He was only 58 years old. Geschwind is now recognised as one of most influential figures in post-war neuroscience, who in his many journal articles and books left a legacy of knowledge and ideas that led to a rebirth of what he termed behavioural neurology (an alternative term might be clinical neuropsychology). Although his theory of disconnection syndromes would come under some criticism in the 1970s, when researchers such as Marsel Mesulam and Antonio Damasio showed that certain association areas of the brain had their own specific behavioural functions, the recent development of non-invasive imaging techniques such as diffusion tensor imaging (an MRI-based neuroimaging technique that makes it possible to estimate the location and orientation of the brain's white matter tracts) has helped reinvigorate interest in Geschwind's connectionist theory of brain function and its clinical importance. Indeed, as we learn more about the brain's connectivity (a subject now very much in vogue) it is likely that Geschwind's approach to neuropsychology will be vindicated yet further.

Geschwind's major writings

Geschwind, N. (1965) Disconnexion syndromes in animals and man. I. *Brain*, 88, 237–294.

Geschwind, N. (1965) Disconnexion syndromes in animals and man. II. *Brain*, 88, 585–644.

Geschwind, N. and Levitsky, W. (1968) Human brain: Left-right asymmetries in temporal speech region. *Science*, 161, 186–187.

Geschwind, N. (1972) Language and the brain. *Scientific American*, April, 226, 76–83.

Geschwind, N. and Behan, P. (1982) Left-handedness: Association with immune disease, migraine and developmental learning disorders. *Proceedings of the National Academy of Sciences*, 79, 5097–5100.

Geschwind, N. and Galaburda, G.M. (1984) *Cerebral Dominance: The Biological Foundations*. Boston: Harvard University Press.

Geschwind, N. and Galaburda, G.M. (2003) *Cerebral Lateralization: Biological Mechanisms, Associations and Pathology*. Cambridge, MA: MIT Press.

Further reading

Absher, J.R. and Benson, D.F. (1993) Disconnection syndromes: An overview of Geschwind's contributions. *Neurology*, 43, 862–867.

Catani, M. and Ffytche, D.H. (2005) The rises and falls of disconnection syndromes. *Brain*, 128, 2224–2239.

Catani, M. and Mesulam, M.M. (2008) What is a disconnection syndrome? *Cortex*, 44, 911–913.

Damasio, A.R. (1985) Norman Geschwind (1926–1984). *Trends in Neurosciences*, 8, 388–391.

Devinsky, O. and Schachter, S.C. (1993) *Norman Geschwind: Selected Publications on Language, Epilepsy and Behavior*. Oxford: Butterworth-Heinemann.

Mesulam, M.M. (2015) Fifty years of Disconnexion syndromes and the Geschwind legacy. *Brain*, 138, 2791–2799.

Schachter, S.C. and Devinsky, O. (1997) *Behavioral Neurology and the Legacy of Norman Geschwind*. New York: Lippincott-Raven.

Chapter 26

Paul Greengard (b. 1925)

Greengard has shown that many types of neurotransmitters in the brain exert their physiological effects by stimulating chemicals called "second messengers", which cause the phosphorylation of key proteins involved in a wide range of intracellular processes.

Paul Greengard was born in New York City. His father was of Baltic German ancestry and had been a relatively successful vaudeville singer and dancer as a young man, although he later turned to business in the cosmetics field. Sadly, Greengard's Jewish mother died giving birth to her son – resulting in his father marrying again just over a year later, to an Episcopalian who raised her stepson as a Christian. Although Greengard went to public schools in Brooklyn and Queens, before finishing high school education at the age of 17, he was discouraged from going to college by his abusive father and therefore joined the Navy as an electronics technician to serve in World War II. Seeing 3 years of service, Greengard spent some of his time assigned to a team at the Massachusetts Institute of Technology involved in developing an early-warning system to intercept Japanese kamikaze planes. After the war, Greengard attended Hamilton College in upstate New York, where he graduated in mathematics and physics in 1948. His intention was then to study theoretical physics, but the only chance of a fellowship at this time was from the Atomic Energy Commission. This was just 3 years after the atomic bombs had been dropped over Japan, and Greengard did not want to contribute to producing more weapons of mass destruction. Consequently, Greengard decided to pursue medical physics, an area still very much in its infancy, at the University of Pennsylvania. It was here that Greengard was first introduced to neurophysiology, although he would leave after a year to join the new department of biophysics at John Hopkins, where he undertook his first research under the neurophysiologist H.K. Hartline.[1]

In the second year of his studies, Greengard began developing an interest in neurochemistry and cell biology. The impetus for this change came after Greengard heard a talk by Alan Hodgkin describing his work (with Andrew Huxley) that explained the ionic basis of the nerve impulse. Following this, Greengard was convinced that there was little more to discover in neurophysiology – and he started working on the molecular and biochemical basis of nerve function instead. After obtaining his PhD in 1953, Greengard went to Europe, spending time at Maudsley Hospital in London, Cambridge and Amsterdam, before settling back in London where he worked in the laboratory of Wilhelm Feldberg at the NIMR.[2] Greengard considered staying in England, a country he liked, but the lack of equipment (and central heating) led him back to the United States in 1958, where he took up directorship of the biochemistry department for the drug company Geigy in New York. Concerned primarily with the development of antidepressant drugs, Greengard eventually grew frustrated with the lack of innovative research projects and left in 1967. He then took a 1 year sabbatical, working with Alfred Gilman in New York and Earl Sutherland at Vanderbilt University, before joining Yale as full professor of pharmacology in 1968.

Greengard's time with Sutherland would prove particularly inspiring and strongly influence the direction of his research thereafter. In the mid-1950s, Sutherland had examined how the pancreatic hormone glucagon breaks down glycogen in the liver to provide glucose for energy. By homogenising the liver tissue into separate fractions, he discovered these hormones activated a class of enzymes known as the adenylate cyclases. But more importantly, Sutherland found these enzymes created cyclic adenosine monophosphate (cAMP). This mimicked the action of glucagon, causing the breakdown of glycogen into glucose. It soon became apparent that the hormone glucagon acted as a *first messenger* in this process that led to the formation of the *second messenger* cAMP which initiated the chemical events inside the cell. The discovery of second messengers would win Sutherland the Nobel Prize in 1971. Around the same time, Edwin Krebs at Washington State University in Seattle showed that adrenaline had a similar effect on the breakdown of glycogen into glucose via its stimulation of cAMP – but this time in skeletal muscle. Krebs also identified the next step in the chain of intracellular chemical events when he realised that cAMP phosphorylates[3] an enzyme called "phosphorylase kinase",[4] which then phosphorylates a second enzyme to break down glycogen into glucose. Krebs won the Nobel Prize for this discovery in 1992.

The discovery of second messengers was a major breakthrough in the field of biochemistry, opening up numerous new avenues of research on how cells process information after they have been stimulated by hormones (or other chemical substances). Indeed, a number of different second messengers have now been identified (e.g. cyclic guanosine monophosphate [cGMP], calcium, inositol triphosphate, diacylglycerol, etc.), and these are found in cells throughout the body, where they provide a type of biological amplification. In effect, a first messenger will often induce the formation of many molecules of a second messenger, setting off a blast of biochemical chain reactions inside the cell. In turn, this type of amplifying signal has been shown to be involved in a huge range of intracellular biochemical processes and effects.

Greengard was inspired by Sutherland's research to undertake the highly ambitious task of looking for second messengers in the nervous system and establishing whether cAMP played a role in neurotransmitter action. One of the first systems Greengard examined was the superior cervical ganglion (SCG). This is a collection of sympathetic ganglia that arises from the preganglionic fibres of the upper four cervical nerves, which come together to form a large ganglion, whose postganglionic fibres then branch out to innervate parts of the head, including the pupils of the eye, salivary glands and blood vessels. Greengard found that preganglionic stimulation increased the levels of cAMP in the SCG, whereas postganglionic stimulation did not. But more importantly, Greengard found that the application of dopamine to the SCG also increased levels of cAMP, which mimicked the effects of preganglionic electrical stimulation. The implication was clear: the neurotransmitter dopamine appeared to produce its effects through the intracellular formation of cAMP.

Of interest to Greengard was the fact that the postsynaptic neurons of the SCG also exhibited two types of response to preganglionic stimulation – fast and slow. As the name suggests, fast responses took place very rapidly (in about one-thousandth of a second) and produced a transient electrical potential in the postganglionic neuron lasting no more than a few milliseconds. In contrast, the slow response took about 200 to 300 msecs to form and could last up to 2 minutes. Greengard found that dopamine and its associated cAMP activity was only associated with the slow response. The concept of slow synaptic neurotransmission would play a pivotal part in Greengard's theorising over the subsequent years.

The next obvious step was to look for cAMP in the brain. Greengard's initial studies were highly encouraging since he not only found cAMP-dependent protein kinase activity in the bovine brain but also found the concentration of this enzyme to be much higher than that in the liver. Even more exciting were the results of subcellular fractionation studies showing the enzyme to be confined to the synaptic region of nerve cells – data consistent with a possible role for cAMP in synaptic transmission. Indeed, this was proven to be the case in 1971 when Greengard and his graduate student John Kebabian reported adenylate cyclase activity in tissue extracted from the striatum exposed to dopamine. It was the first demonstration of neurotransmitter-linked intracellular second messenger activity in the nervous system

Greengard's method of using striatal tissue homogenates to test for cAMP activity also lent itself nicely to testing another important theory. In 1963, Carlsson and Lindqvist had proposed that the antipsychotic drugs chlorpromazine and haloperidol worked by blocking dopamine receptors (see Chapter 24). However, there was no effective way of testing this theory at the time. But Greengard's new method provided a test: if chlorpromazine and haloperidol were added to striatal homogenates and found to inhibit dopamine's action on cAMP, it would support the antipsychotic blocking hypothesis. This was duly proven to be the case (Kebabian et al. 1972). Later research by Kebabian would also show that there were different types of dopamine receptors (D-1 and D-2) linked to adenylate in different ways.

Soon after Greengard's laboratory had established the role of cAMP in several types of neural functioning, another substance was recognised as having second messenger actions in the brain and some other tissues. This was cGMP, which was shown to regulate a different type of kinase inside the cell (Kuo and Greengard 1970). And in 1979, Greengard's group outlined a third type of second messenger system, stimulated by calcium, which formed a substance called "calmodulin" (an abbreviation for calcium-modulated protein) that regulated many different protein targets and cellular functions. In fact, over the next few years, Greengard's laboratory would find more than 100 proteins for protein kinases to act upon, which were highly enriched or exclusively localised in the brain. They also showed that these proteins mimicked or antagonised the abilities of neurotransmitters to produce responses inside nerve cells. In short, they provided overwhelming evidence of a role for second messengers and signalling molecules in synaptic transmission.

One group of intracellular proteins discovered by Greengard which are subject to phosphorylation are called the synapsins. In fact, these are now recognised as the most ubiquitous group of proteins in the brain, found in the nerve terminals, where they interact with the vesicle membranes that store the chemical transmitters. Greengard has shown that among their many functions, synapsins are essential for maintaining the pool of vesicles and enhancing (through phosphorylation) or inhibiting (though de-phosphorylation) the release of neurotransmitters. But arguably the most important protein discovered in Greengard's laboratory is called "dopamine and cAMP-regulated phosphoprotein" (DARPP-32). This is a vital regulator of neuronal function that affects a large number of other proteins. In fact, DARPP-32 is now known to integrate input from several neurotransmitters, including glutamate, and is found throughout the brain – although it is particularly rich in the dopamine-rich striatum and nucleus accumbens. DARPP-32 has been shown to affect the phosphorylation of a large array of intracellular proteins including those involved in voltage-gated ion channels, ion pumps and gene expression. In addition, a number of drugs have been shown to affect the phosphorylation of DARPP-32, including antipsychotics, and most drugs of abuse such as cocaine, marijuana, nicotine, alcohol and LSD. Interestingly, studies that have used "knockout" mice in which the gene for producing DARPP-32 is missing have also shown that the behavioural effects of fluoxetine (better known as Prozac) are abolished. Thus, DARPP-32 is likely to be an important mediator of antidepressant drugs too. Perhaps most intriguingly, Greengard has also reported that levels of DARRP-32 are reduced in the post-mortem brains of schizophrenic patients. Although this need not mean that DARPP-32 is a cause of schizophrenia, it does seem to play a significant role in the signalling pathways that affect the disease's course and as such could be a future therapeutic target.

Another legacy left to us by Greengard is the idea that nerve cells communicate with each other by two mechanisms: fast and slow synaptic transmission. Fast acting neurotransmitters, which include the excitatory glutamate and the inhibitory GABA, achieve their effects on the target cell in one-thousandth of a second by opening ligand-operated ion channels. Slow-acting neurotransmitters, such as dopamine and serotonin, exert their actions over many seconds or minutes by a more complex sequence of biochemical steps involving second messengers, protein kinases and protein phosphorylation. Slow-acting transmitters also appear to enhance the efficiency by which fast-acting transmitters can act on postsynaptic receptors. Greengard has also suggested that fast synaptic transmission can be considered to represent the hardware of the brain and slow synaptic transmission its software – an idea that may one day have profound implications for neuroscience.

In 1983, Greengard, who is married to the artist Ursula von Rydingsvard, moved from Yale to Rockefeller University to head the Laboratory of Molecular and Clinical Neuroscience. Greengard has remained there since, in charge of over sixty staff, where he works 6 days a week, despite being in his nineties. In 2000, he was awarded the Nobel Prize for his discoveries concerning "signal transduction in the nervous system", and with the winnings, he set up an annual award to encourage women to take up biological research, named in honour of his mother (The Pearl Meister Greengard Prize). Greengard shows no signs of slowing down, and when asked about his commitment to work, he replied "It's more fun understanding the brain than doing a crossword puzzle. It's the same kind of cognitive challenge, but it's much more fun and you feel you're doing something worthwhile". At the time of writing, Greengard had published close to a thousand important research papers – a truly remarkable record.

Notes

1 H.K. Hartline was a co-recipient of the 1967 Nobel Prize in physiology or medicine for his work in analysing the neurophysiological mechanisms of vision.
2 See the chapter on Henry Dale.
3 After being phosphorylated, the protein (or enzyme) assumes a different three-dimensional (3-D) configuration, which also alters its function.
4 A kinase is an enzyme that modifies proteins by adding phosphate groups in a process called "phosphorylation". It is now recognised

that phosphorylation is a general process affecting countless cellular proteins.

Greengard's major writings

Fienberg, A.A. and Greengard, P. (2000) The DARPP-32 knockout mice. *Brain Research Reviews*, 31, 313–319.
Greengard, P. (1978) Phosphorylated proteins as physiological effectors. *Science*, 199, 146–152.
Greengard, P., Allen, P.B. and Nairn, A.C. (1999) Beyond the dopamine receptor: The DARPP-32/protein phosphatise-1 cascade. *Neuron*, 23, 435–447.
Greengard, P. (2001) The neurobiology of slow synaptic transmission. *Science*, 294, 1024–1030.
Greengard, P. (2003) The neurobiology of dopamine signalling. In: H. Jörnvall (ed.), *Nobel Lectures, Physiology or Medicine 1996–2000*. Singapore: World Scientific Publishing Co.
Greengard, P. and Nestler, E.J. (2013) A conversation with Paul Greengard. *Annual Review of Pharmacology and Toxicology*, 53, 1–16.
Kebabian, J.W. and Greengard, P. (1971) Dopamine-adenyl cyclase: possible role in synaptic transmission. *Science*, 174, 1346–1349.
Kebabian, J.W., Petzold, G.L. and Greengard, P. (1972) Dopamine-sensitive adenylate cyclase in caudate nucleus of rat brain, and its similarity to the "dopamine receptor". *Proceedings of the National Academy of Sciences*, 60, 2145–2149.
Kuo, J.F. and Greengard, P. (1970) Cyclic nucleotide-dependent protein kinases. *The Journal of Biological Chemistry*, 245, 2493–2498.
Svenningsson, P., Nishi, A., Fisone, G., Girault, J.A Nain, A.C. and Greengard, P. (2004) DARPP-32: an integrator of neurotransmission. *Annual Review of Pharmacology and Toxicology*, 44, 269–296.

Further reading

Hancock, J.T. (2016) *Cell Signalling*. Oxford University Press: Oxford.
Moyer, P. (2002) Paul Greengard, PhD: On the journey to acceptance of slow synaptic transmission. *Neurology Today*, 2, 16–17.
Snyder, S.H. (2011) What dopamine does in the brain. *Proceedings of the National Academy of Science USA*, 108, 18869–18871.
Ravindran, S. (2011) Paul Greengard: Signals underlying moods, addictions, and brain disorders. *Proceedings of the National Academy of Science USA*, 108, 18872–18874.
Rowland, L.P. (2003) Paul Greengard. In: *NINDS at 50*. New York: Demos Medical Publishing.

Chapter 27

Benjamin Libet (1916–2007)

Although Libet is a relatively unknown neuroscientist, he undertook pioneering and thought-provoking research into the neural basis of human consciousness, initiation of action and free will – areas that also share an overlap with the concerns of philosophy.

Benjamin Libetsky (later Libet)[1] was born in Chicago and was the first child of young Jewish immigrants. His father emigrated to the United States in 1909, from a town called Brusilov in the Ukraine, and found work as a machine operator in a clothing factory. Libet's mother was an émigré from Kiev in 1913 and had been a school teacher. Although Libet's first language was Yiddish, he would soon learn English on the streets of the west side of Chicago where he lived. Despite this unpromising start, Libet enjoyed school and was encouraged by his mother to do well. Libet was also blessed with a powerful alto voice, and when he was 9 years old, he passed an audition to sing in a choir backing the world-renowned cantor Josef Rosenblatt who performed in Chicago. He not only won $25 but was also able to spend some of the proceeds on a piano. Libet continued singing in professional choirs and synagogues, which provided him a useful source of income well into his twenties. Leaving state education at the age of 18, Libet won a half-tuition scholarship to the University of Chicago after undertaking a 3-hour competitive examination in a subject of his choice (chemistry). His high school then agreed to finance the other half of his degree. This was a 4-year course, and after his second year, Libet decided to major in physiology after coming under the influence of renowned neurophysiologist Ralph Gerard. Graduating with a BSc in 1936, Libet was set for medical school, although he changed his mind when deciding a university career provided a more attractive option. Thus, he enrolled for a PhD under the supervision of Gerard, who set him the task of exploring the electrical activity of the frog's brain – which remarkably still exhibited electrical activity when removed from the body and placed into a dish. This would lead to a notable discovery when Libet found that caffeine-induced EEG waves could travel across a complete transaction of the frog brain – demonstrating that the spread of extracellular fields of electric current flow was in itself a possible means of neural interaction and synchronisation. Sadly, Libet was unable to continue with this research after he became allergic to frog's blood, which gave him severe bronchial asthma.

Libet obtained his PhD in 1939 at a relatively young age of 23. However, for the next 15 years or so, his academic life would be one of an itinerant neuroscientist with a number of different positions and research interests. His first post was as an instructor in physiology at the University of Albany in New York State, although this would not last long due to the outbreak of World War II. In 1943, Libet moved to the University of Pennsylvania where he taught medical students while inventing a face mask with an inbuilt oxygen bottle for aviators shot down over the sea. Soon after, Libet would work on an army base near Dayton, Ohio, where he designed and tested clothing for the Air force. Returning to the University of Chicago in 1945, Libet focused his interest on the study of slow potentials in the cerebral cortex (these are essentially replicable current shifts in the EEG that can take from 500 msecs to over several seconds to appear after stimulus onset). Unfortunately, the bitterly cold Chicago winters worsened Libet's asthma, and in some desperation, he sought employment in California. He first took a post in Vallejo as director of a neuromuscular rehabilitation unit and then as assistant professor at the University of California in San Francisco (UCSF) in 1949 – which was much more to his liking. These were difficult times for Libet, and only 2 months after joining UCSF, he nearly lost his life when a duodenal ulcer broke and severed a small artery. He only survived after a 14 pint blood transfusion. Over the course of the next decade or so, much of Libet's research activity would be spent in the Biomechanics Laboratory where he became interested in the neurophysiological properties of various ganglia preparations using intracellular recording. However, he did manage a sabbatical year in 1956 when he worked with John Eccles in Canberra, Australia.

One of the researchers Libet became acquainted with at UCSF was the experimental neurologist Bertram Feinstein who had qualified as a neurosurgeon in the early 1950s. In 1958, Libet received an unexpected offer from Feinstein, who by then was using stereotaxic surgery to implant lesioning electrodes in patients with Parkinson's disease and cerebral palsy. He asked Libet if he would like to explore the brains of his fully conscious patients for a short period prior to their surgery. Not surprisingly, Libet took up this "wonderful opportunity" and over the next 8 years he would perform studies on more than a 100 patients, all of whom gave their consent along with the approval of the hospital committee. Feinstein's surgery required the skull to be opened prior to the operation (a procedure only requiring a local anaesthetic to the scalp and the periosteal tissue covering the bone), which exposed parts of the frontal and parietal lobes. Libet decided to examine the somatosensory cortex lying in the postcentral gyrus of the parietal lobes, which he could easily access.

It had been known since the work of Penfield in the 1930s that an electrical stimulus applied to the somatosensory cortex could induce a touch sensation that appeared to arise from the corresponding part of the skin or body. In fact, Penfield had shown that the somatosensory cortex has a topographical or map-like relationship with the body – so the brain areas representing the legs and feet are found in the gyrus near the top of the head and the face and head at its lowest end. Hence, Libet would have known the likely effects of stimulation, or what parts of the body were likely to be "felt" after stimulation. However, Libet began his investigations by trying to establish the barest threshold parameters needed to produce a conscious reaction in his subjects – and to this end he stimulated their somatosensory cortex with electrodes placed on its surface and varied the frequency, duration and intensity of the electrical pulses. This led Libet to discover that the most important determinant of producing a conscious touch sensation was the duration of stimulation. That is, when Libet stimulated the cortex with brief pulses of small amplitude current for periods up to about 500 msecs, the subjects reported nothing. But when he went beyond a stimulation period of 500 msecs (or half a second), the subjects typically reported a sensation such as a touch on the skin. These results were intriguing: they implied that at least half a second of brain processing was necessary for a simple conscious percept to emerge from the stimulation.

One possible objection to this conclusion is that direct electrical stimulation of the brain is a highly artificial procedure. Hence, a more realistic test of Libet's hypothesis came when he ran a similar type of experiment, but this time stimulated the medial lemniscus – a large ascending pathway that carries sensation from the skin, joints and muscles to the ventral thalamus, from where it passed to the somatosensory cortex (fortunately for Libet, some of Feinstein's patients already had electrodes placed in this pathway). Again, the findings showed that touch sensations were never consciously reported from stimulation of the medial lemniscus, unless they were at least 500 msecs in duration. In Libet's terminology, this was the amount of time needed for *neuronal adequacy* to produce a conscious sensation. Interestingly, this seemed to contradict the fact that a brief stimulus of just a few milliseconds applied to the skin can often be consciously perceived. But in further studies, Libet showed that even a brief "touch" stimulus of a few milliseconds resulted in the occurrence of evoked potentials in the somatosensory cortex that exceeded the 500 msec threshold.

The most startling implication of Libet's findings is that our conscious perception and experience of the real world lags half a second behind the real events that are actually taking place. But can this really be so? According to many psychologists, we would surely notice the delay between stimulus and reaction. However, to explain our seamless perception of the world, Libet came up with a controversial proposal: he argued that once neuronal adequacy has occurred, our sensory experiences are then subjectively referred back in time. Thus, when we consciously perceive a stimulus after 500 msecs of neural processing in the brain, at that point the experience is referred back to the actual time when it happened. Although this sounds implausible, Libet had a neat experiment to back up his theory. He simply asked his subjects to report what of two sensations came first: an ordinary touch of the skin or an electrical stimulus applied to the somatosensory cortex. Libet found that the subjects always reported the touch sensation from the skin first, even though this might arrive in the brain some 450 msecs after the neural stimulation of the cortex had begun. In other words, the conscious awareness of touch always came near the end of the stimulation period needed for neuronal adequacy to occur in the cortex – exactly what is predicted by the referral theory.

In 1977, Libet turned his attention to another issue concerning the timing of consciousness: namely, how long does it take one to make a decision to perform an intentional act? Libet investigated this by using an EEG measure known as the "readiness potential", which had been first discovered by Hans Helmut Kornhuber and Lűder Deecke in 1965. These researchers had found that when their subjects were asked to voluntarily move their wrist or fingers, a highly distinctive slow electrical change in brain activity could be recorded from the upper part of the head, which took place about 800 msecs before any movement took place. In fact, the readiness potential is now recognised as a feature of the brain's motor cortex and supplementary motor area, which always takes place prior to an action. Libet wanted to see if the readiness potential also preceded the conscious decision to act, and to do this, he set up what is arguably his most famous and controversial experiment. To solve his problem, Libet had to measure three events: (1) the start of the voluntary action itself, (2) the start of the readiness potential and (3) the moment of the conscious decision to act. The first two are easy enough to time (e.g. by noting the start of the potential and using electrodes on the appropriate muscle), but the third measure, being subjective, was trickier. Nonetheless, Libet came up with a clever solution. The method, put simply, was to have a subject observe the equivalent of a clock and then when they became aware of an intention to "act", to report that "clock time" to the experimenter. In fact, the procedure was more intricate than this, since the clock was presented on a screen, and had a spot of light revolving at a speed twenty-five times faster than that of a normal sweep hand. Each second indicated on the clock screen, therefore, was equivalent to about 40 msecs. In one condition, subjects could flex their wrist at any chosen time but had to maintain their gaze at the clock, and note the position of the light spot when they became aware of the intention to move. Libet first tried this technique in 1978 and found to his surprise that subjects could report these timings with a reliability of ± 20 msecs. He also addressed a number of other issues concerned with the method's validity.

The common sense view is that the conscious intent to act should manifest itself before, or at the beginning, of the readiness potential. But Libet showed this was not the case. In fact, the conscious decision to act took place approximately 400 to 350 msecs after the onset of the readiness potential. In other words, the voluntary action began unconsciously almost half a second before the conscious will to move. These findings are truly startling for they imply our minds are made up unconsciously, long before our conscious awareness has had time to form and act. They can also be used to show that free will is an illusion – and some researchers have come to this conclusion. Indeed, one such psychologist, Daniel Wegner, has argued that consciousness is an *epiphenomenon* – that is, something caused by brain events but which does not have any influence on our behaviour. Libet, however, believed otherwise. He pointed out that while the readiness potential precedes the conscious intention to act, the interval between intent and muscular activation (which is about 200–150 msecs) gives enough time for consciousness to decide whether to go ahead with the action or abort it – an essential function of awareness that Libet called *the veto*. More generally, this veto allows us to be responsible for such acts as murdering, raping or stealing from people – because we can, and must be, held responsible for stopping ourselves from doing those things.

Libet's work on consciousness and free will is some of the most controversial in this book, and it should be noted that there are many other interpretations of his work. Nonetheless, there is little doubt that Libet's work has helped make the study of consciousness and free will more scientifically respectable – and many of the questions he has raised remain important and largely unanswered. Indeed, one might go as far to say that Libet's approach is just about the only one yet to yield any credible evidence about how the brain produces conscious awareness. After working at UCSF for 35 years, Libet retired in 1984, although he continued with research and writing – most notably publishing *Mind Time: The Temporal Factor in Consciousness* in 2004 – a book that not only described his experiments in layman's terms but also extended a theory (first proposed in 1994) that saw consciousness as an emergent mental field, somewhat analogous to electromagnetic and gravitational forces but which cannot be described in terms of any observable constituents. Nonetheless, if his theory is correct, then Libet argues it should be possible for isolated slabs of cortex to communicate consciously with each other through force fields (page 178). Needless to say, it is an idea that is impossible (and unethical) to test. Libet died in 2007 at the age of 91.

Note

1 When faced with the publication of his first scientific paper in 1938, which was also at the time of the Great Depression and anti-Jewish

bias in universities, Benjamin Libitsky changed his name to Libet to make it sound more American. Libet liked the change because it made him sound French!

Libet's major writings

Libet, B. (1985) Unconscious cerebral initiative and the role of the conscious will in voluntary action. *Behavioral and Brain Sciences*, 89, 567–615.

Libet, B. (1996) Benjamin Libet. In: L.R. Squire (ed.), *The History of Neuroscience in Autobiography, Volume 1.* New York: Elsevier Academic Press.

Libet, B. (1999) Do we have free will? *Journal of Consciousness Studies*, 6, 47–57.

Libet, B., Freeman, A. and Sutherland, K. (1999) *The Volitional Brain: Towards a Neuroscience of Free Will.* Exeter: Imprint Academic Press.

Libet, B. (2004) *Mind Time: The Temporal Factor in Consciousness.* Cambridge, MA: Harvard University Press.

Further reading

Blackmore, S. (2003) *Consciousness: An Introduction.* London: Hodder and Stoughton.

Gomes, G. (1998) The timing of conscious experience: A critical review and reinterpretation of Libet's research. *Consciousness and Cognition*, 7, 559–595.

Klein, S. (2002) Libet's timing of mental events: Commentary on the commentaries. *Consciousness and Cognition*, 11, 326–333.

Obhi, S.S. and Haggard, P. (2004) Free will and free won't. *American Scientist*, 92, 358–365.

Spence, S.A. (2009) *The Actor's Brain: Exploring the Cognitive Neuroscience of Free Will.* Oxford: Oxford University Press.

Wegner, D.M. (2002) *The Illusion of Conscious Will.* Cambridge, MA: MIT Press.

Chapter 28

Eric Kandel (b. 1929)

Eric Kandel is famous for introducing the humble sea slug Aplysia into neuroscience and using a variety of approaches, from behavioural conditioning to molecular analysis, to show that learning and memory are dependent on the efficacy of synapses that can be modified in many different ways.

Eric Kandel was born in Vienna to Jewish parents who had come from the Ukraine. These were difficult times since the Austro-Hungarian Empire had largely disintegrated after its defeat in World War I, and despite Kandel's father opening a toy store shortly after his birth, the family were forced to leave Vienna due to anti-Semitism after Austria was annexed by Germany. Consequently, when Kandel was just 9 years old, he and his brother boarded the *S.S. Gerolstein* at Antwerp and sailed to join their uncle in Brooklyn, soon followed by his parents. After settling in New York, where his father worked in a toothbrush factory, Kandel won a scholarship to a prestigious private Jewish school after being tutored in Judaic studies by his grandfather. Completing his basic schooling at the age of 14, Kandel entered Erasmus Hall High School where he became interested in history, literature and girls. He also worked on the school's newspaper and became its sports editor. This does not appear to have distracted him from his academic work for at the insistence of his history teacher, Kandel applied to Harvard and was only one of two students out of a class of 1,400 to win a scholarship. Here, in 1951, Kandel would graduate in European history and literature. But more importantly, Kandel had also been introduced to psychoanalysis by a fellow student and became so enthusiastic with this new study of the mind that he took extra courses enabling him to become a medical student. His ambition was to become a psychiatrist. Entering medical school at New York University in 1952, Kandel's interest in psychoanalysis led him to discover more about the brain's biology, and in 1955, he began working in the laboratory of neurophysiologist Harry Grundfest at Columbia (Grundfest was best known for his work showing that the velocity of the action potential was dependent on the neuron's axon diameter – with bigger axonal diameters producing faster nerve impulses). Kandel greatly enjoyed his new research experience, describing it as "doing experiments, not having to wear a neck tie and laughing a lot". It was also during this time he first met Stephen Kuffler, a man Kandel referred to as one of his neurobiological heroes and pioneer of the first intracellular recordings of sensory nerve cells isolated from lobster and crayfish. Determined to learn this technique, Kandel would spend some of his internship in 1956 gaining experience in making microelectrodes and recording from the crayfish giant axon. In 1957, Kandel was nominated to join the NIH in Bethesda – an alternative to serving in the physician's draft for military service.

By now, Kandel was focusing his attention on the biological basis of learning and memory in higher animals. Kandel had first become drawn to this subject while at Harvard where the eminent psychologist and learning theorist B.F. Skinner was professor. But a greater influence was the paper published in 1957 by William Scoville and Brenda Milner describing the case of HM who suffered profound anterograde amnesia as a result of bilateral medial temporal lobe excision that had included much of his hippocampus. At the time, nothing was known about the cellular or molecular mechanisms of memory, and Kandel reasoned it would be useful to apply his newly acquired intracellular recording techniques to explore the characteristics of pyramidal cells in the hippocampus. Kandel was helped in this quest by the young medical graduate, Alden Spencer, and they obtained some interesting results using anaesthetised cats as subjects. They showed, for example, that action potentials could be generated in the dendritic trees of hippocampal pyramidal cells and not just at the axon hillock as was widely believed. In addition, they found pyramidal neurons had a spontaneous rhythmical activity, as well as exerting a strong inhibitory influence over other cells. Although these were significant findings, Kandel recognised that they did not explain how the hippocampus stores and consolidates memory. Indeed, he suspected the site of memory lay in the synaptic connections between neurons

– a theory that at the time was impossible to investigate with cellular recording of the hippocampus.

One step towards solving the problem was to examine learning and memory in a simpler structure than the hippocampus, and towards this end, Kandel began searching for a more suitable preparation. Kandel considered using crayfish, lobster, flies and even nematode worms, but instead settled on the giant marine snail *Aplysia californica*, a creature that can grow up to 30 cm in length and weigh 2 kg. It proved to be a wise decision for several reasons. First, *Aplysia's* nervous system is relatively simple compared to that of vertebrates, with approximately 20,000 neurons organised into nine different ganglia. Yet, the biophysical and biochemical properties of its neurons are the same as those of other creatures. Second, the neurons of *Aplysia* are large and can be seen with a light microscope. Third, the neurons making up the nervous system are fixed, or found in the same location, making them easy to identify. But perhaps most importantly, *Aplysia* had a rich repertoire of behaviours and exhibited several types of simple learning. Nonetheless, many still remained sceptical over Kandel's chances of success with his new preparation.

In 1962, Kandel managed to get a postdoctoral fellowship to work for 16 months in the Paris laboratory of Ladislav Tauc, who at the time was one of only two people in the world working with *Aplysia*. Here, Kandel became familiar with the gill withdrawal reflex. He also began stimulating and recording from neurons in *Aplysia's* abdominal ganglia – revealing an effect called heterosynpatic facilitation. This is where stimulation of two separate sensory axons entering the ganglion (which simulates a conditioning procedure) was able to cause a long-lasting increase in excitatory activity in a recipient motor cell, which lasted for over 30 minutes. Kandel also knew where the synapses were located in this system, and he reasoned that long-lasting chemical neurotransmission was responsible for the enhancement. In physiological terms at least, Kandel had demonstrated learning in *Aplysia* and found a neural system that he could explore further. These initial findings were published in 1965.

In 1965, Kandel took up a position in the medical school at NYU, where he set up a small research group – a place where he would remain until 1974 when he became founding director of the Center for Neurobiology and Behavior at Columbia University. Here, Kandel began experiments with several colleagues on the behavioural manifestations of learning in intact *Aplysia*. In particular, Kandel focussed on habituation of the gill withdrawal reflex. *Aplysia* has a large gill located on its back, which is responsible for extracting oxygen from water, and if its exhalent siphon is touched, the animal will vigorously retract the organ into a protective cavity that is then covered by a large fleshy pad called the "mantle". However, if a weak stimulus is applied repeatedly to the siphon, this response will lessen or habituate, and eventually not occur at all. In other words, the slug has "learned" that the stimulus is harmless. Despite this, if after habituation, *Aplysia* receives a strong stimulus to the tail, the gill withdrawal reflex will be restored – a process Kandel called sensitisation.

But to fully understand this process at a cellular level, it was first necessary to identify the neural circuitry underlying the reflex. To do this, Kandel, along with several others including Irving Kupfermann, made a slit in *Aplysia's* neck, enabling the ganglionic nerve pathways to be exposed – following which they were able to stimulate and record from various neurons to mimic the reflex. They showed (at its most simple) that the siphon contained a total of twenty-four sensory neurons that were sensitive to tactile information, whose axons projected onto a cluster of six motor neurons that controlled the contraction of the gill. Once this had been established, Kandel and Kupfermann were then able to establish where the synaptic changes mediating the changes in the gill reflex were taking place. The answer was as Kandel had anticipated: each tactile stimulus was causing a decrease in the amount of neurotransmitter (which was later shown to be glutamate) from the axon terminals of the sensory neurons. The reduced flow of transmitter then produced less stimulation of the motor neurons innervating the gill, thereby causing the habituation. In the case of sensitisation, however, the release of neurotransmitter was increased. These were significant findings for they were the first to show that the strength and effectiveness of synaptic connections could be altered by experience – work that would be published in three consecutive papers in *Science* in 1970.

Soon after these studies, Kandel, along with biochemist James Schwartz, began focusing on identifying the chemical changes within the neurons themselves, which underpinned *Aplysia's* ability to learn. In their first study (1971), they found that blocking protein synthesis for 24 hours did not prevent the short-term synaptic changes associated with habituation and sensitisation. Thus, no permanent structural change appeared to be taking place in the synapse. Kandel and Schwartz then began to suspect that the short-term changes responsible for learning might involve intracellular biochemical systems instead. This was confirmed when Kandel injected the second messenger cAMP into the sensory neurons and found that it mimicked the sensitising stimulus by enhancing neurotransmitter (glutamate) release. It was a highly significant finding, which according to Kandel "gave us our first insight into the molecular mechanisms of short-term memory". Soon after, the neurotransmitter serotonin was found to be released by the interneurons mediating the sensitisation response.

Another important question concerned the process by which cAMP enhanced glutamate release. It soon became clear that cAMP did this by reducing the flow of potassium into the cell, which caused more calcium to enter the presynaptic terminal with each action potential. But how did cAMP reduce the flow of potassium into the cell? To answer this, Kandel, in collaboration with Paul Greengard, showed that cAMP stimulated an enzyme known as protein kinase A (PKA) that added phosphate groups to the potassium channel, thereby changing its conductance (Castellucci et al. 1980). This molecular change enhanced transmitter release, while an inhibitor of PKA was shown to block this cAMP-mediated effect.

Up to this point, Kandel and his colleagues had only examined temporary changes in *Aplysia*, which lasted a few minutes. But the motor neurons of *Aplysia* were also capable of showing long-lasting changes in neurotransmitter release, sometimes taking place over weeks, especially when the snail had been trained with habituation or classical conditioning regimes. This implied that protein synthesis was involved after all. In 1983, Kandel helped create the Howard Hughes Medical Research Institute at Columbia, which was devoted to molecular neural science – after which he began his search for the genes and proteins they synthesised, which were responsible for more permanent synaptic changes in *Aplysia*. But to do this, Kandel had to first develop a tissue culture system whereby isolated neurons from *Aplysia* could be experimentally manipulated *in vitro* (i.e. in glass dishes). The development of this culture technique, however, soon provided some major discoveries. For example, when a motor neuron from *Aplysia* was repeatedly stimulated by serotonin, Kandel found that a subunit of protein kinase moved into the nucleus where the DNA was stored. Following this, it was shown that one of the nuclear targets for this subunit was a chemical involved in the transcription of DNA called "cAMP response element binding protein" (CREB). This protein bound to certain DNA sequences, which then switched on several immediate-response genes, thereby increasing protein synthesis. Thus, protein synthesis was clearly involved in the synaptic changes underlying long-term learning. In fact, the CREB system became even more complicated with the discovery of CREB-2 that inhibited the effects of CREB-1, so that the final result depended on the balance of the activity of the two CREBs, one a stimulator and the other a repressor.

To complicate the picture further, several other transcription factors have been found in *Aplysia's* neurons – including a class that has been called *Aplysia cell adhesion molecules*. These set in motion a type of protein synthesis that is involved in the production of new synapses. Kandel reports that a typical sensory neuron has about 1200 synaptic connections, although this can rise to about 2600 after long-term sensitisation. However, with recovery time, the number of connections will return to normal.

In 2000, Kandel was awarded the Nobel Prize (along with Arvid Carlsson and Paul Greengard) for his work concerning signal transduction in the nervous system. Over the course of his long career, Kandel has contributed to hundreds of papers that have pioneered and applied neuroanatomical, electrophysiological, behavioural, pharmacological and molecular methodologies to understand the cellular and molecular basis of simple learning. And more recently, he has begun to turn his attention back to the hippocampus. If this productivity wasn't astounding enough, Kandel founded the company Neurogenetics in 1987, which focuses on cloning dopamine and serotonin receptors, and co-founded Memory Pharmaceuticals in 1996 aimed at developing drugs to improve cognition and fight dementia. He is also author of several books including one well known to students entitled *Principles of Neural Science*, first published in 1981, now in its fifth edition. Despite approaching his nineties at the time of writing, Kandel shows no signs of slowing down and remains dedicated to working at Columbia University and the Howard Hughes Medical Institute. Few could have predicted that when Kandel began work on the humble sea slug *Aplysia* in the early 1960s, that it would give rise to so many profound discoveries in neuroscience.

Kandel's major writings

Castellucci, V.F., Kandel, E.R., Schwartz, J.H., Wilson, F.D., Nairn, A.C. and Greengard, P. (1980) Intracellular injection of the catalytic subunit of cyclic AMP-dependent protein kinase stimulates facilitation of transmitter release underlying behavioral Sensitization in Aplysia. *Proceedings of the National Academy of Sciences USA*, 77, 7492–7496.

Kandel, E.R. and Schwartz, J.H. (1982) Molecular biology of learning: Modulation of transmitter release. *Science*, 280, 433–443.

Kandel, E.R. (2000) The molecular biology of memory storage: A dialogue between genes and synapses. In: H. Jörnvall (ed.), *Nobel Lectures, Physiology or Medicine 1996 -2000*. Singapore: World Scientific Publishing Co. (2003).

Kandel, E.R. (2006) *In Search of Memory: The Emergence of a New Science of Mind*. New York: Norton and Co.

Kandel, E.R., Dudai, Y. and Mayford, M.R. (2014) The molecular and systems biology of memory. *Cell*, 157, 163–186.

Kandel, E.R. (2012) The molecular biology of memory: cAMP, PKA, CRE, CREB-1, CREB-2 and CPEB. *Molecular Brain*, 5, 1–12.

Kandel, E.R. (2016) Eric Kandel. In: T.D. Albright and L.R. Squire (eds), *The History of Neuroscience in Autobiography, Volume 9*. Washington, DC: Society for Neuroscience.

Further reading

Bailey, C.H., Kandel, E.R. and Harris, K.M. (2015) Structural components of synaptic plasticity and memory consolidation. *Cold Spring Harbour Perspectives in Biology*, 7, a021758.

Hawkins, R.D., Kandel, E.R. and Bailey, C.R. (2006) Molecular mechanisms of memory storage in *Aplysia*. *The Biological Bulletin*, 210, 174–191.

Kandel, E.R. (1970) Nerve cells and behaviour. *Scientific American*, 223, 57–70.

Kandel, E.R. (1979) Small systems of neurons. *Scientific American*, 241, 67–76.

Kandel, E., Schwartz, J.H. and Jessell, T.M. (2012) *Principles of Neural Science* (5th edn). New York: McGraw Hill.

Mayford, M., Siegelbaum, S.A. and Kandel, E.R. (2017) Synapses and memory storage. *Cold Spring Harbour Perspectives in Biology*, 4, a005751.

Chapter 29

Bert Sakmann (b. 1942) and Erwin Neher (b. 1944)

Sakmann and Neher are famous for developing the patch-clamp method that allowed the tiny currents of single ion channels in cells to be measured in time frames as little as 1 msec – a breakthrough laying the foundation of molecular electrophysiology as a recognised science.

Bert Sakmann was born in Stuttgart, Germany, during World War II, at a time when the city was a major bombing target for the allies. His father was a theatre director and son of a physician whose family had lived in southern Germany for several generations. Sakmann's mother was a physiotherapist, born in Bangkok to a Prussian father who had been a physician to the King of Siam and founder of the first hospital in Thailand. After the war, Sakmann grew up in Lindau on the shores of Lake Constance in Bavaria, where he went to primary and lower secondary schools. In his teens, Sakmann returned to Stuttgart where he completed his gymnasium education (the most advanced of the three types of German secondary school). Sakmann's favourite subject was mechanics – an interest he developed in his spare time by designing and building model motor sailing ships as well as remote control aeroplanes. In terms of a career, Sakmann seemed well suited to engineering, but in the final year of school he came across the subject of cybernetics (the science of self-automated control)[1] and its possible application in biological systems. This fascinated Sakmann because he began to realise that living organisms could be understood in engineering terms. Now recognising the benefits of continuing the family tradition of medicine, Sakmann enrolled at Tübingen University where he studied biochemistry and physiology. He also undertook a thesis in electrophysiology, which utilised some of his engineering skills. A medical degree in Germany took at least 6 years to complete and Sakmann would study at several different medical schools including Freiburg, Berlin and Paris, before completing his medical exams at the University in Munich in 1968. Although he continued to work in Munich as a junior doctor, Sakmann remained interested in cybernetics and became a PhD student under Otto Creutzfeldt at the Max-Planck-Institut für Psychiatrie, where he worked on the neural basis of light adaptation in the cat's visual system. It was here, in 1971, that Sakmann first met Erwin Neher, who taught him the electrophysiological technique of voltage clamping (a method that measures ion flow currents through the neural membrane by setting the cell's voltage at a specified level). Soon after this formative learning experience, Sakmann won a British Council Fellowship to work under Bernard Katz at University College, London. Three years later, in 1974, Sakmann was invited back to Germany to continue his collaboration with Otto Creutzfeldt and Edwin Neher, who were now at the Max Planck Institute for Biophysical Chemistry in Göttingen. In the same year, Sakmann finally completed his PhD.

Erwin Neher grew up in a small town of Buchloe, which lies some 40 miles west of Munich in 1944. His father was involved in the administration of a dairy company that provided vital food supplies during the war, and his mother was a retired teacher who had given up her occupation after getting married to raise a family of five children. It was a fairly affluent family for the times, and Neher spent much of his childhood playing in the big park-like garden of his home where he took an interest in watching nature. After completing his elementary schooling at the age of 10, Neher attended the gymnasium at the nearby town of Mindelheim where he excelled in physics and mathematics. He also eagerly read about cybernetics and anything he could find on the Hodgkin-Huxley theory of the action potential. By the time of his final gymnasium exams in 1963, Neher had settled on a plan to become a biophysist with a view of applying this expertise to biological questions. This led him to enrol on a physics degree course at the University of Munich in 1963, and after graduating in 1963, Neher moved to the University of Wisconsin on a Fulbright Scholarship, where he worked on a project improving the technique of mass spectrometry – research that led to his master of science degree in 1967. Returning to Munich, Neher began his PhD at the Max-Planck-Institut für Psychiatrie with Hans-Dieter Lux, examining synaptic mechanisms and ion currents in snail neurons using the voltage-clamping method. Here, he met Sakmann, and they were to meet again in Göttingen in 1973, where Neher was working on single channel recording in artificial membranes. It was clear that the two men had similar interests and were soon in collaboration, attempting to measure the currents of single ion channels.

The study of ion channels is one of the major endeavours of modern cell biology because they are crucial components in the activity of living cells. Ion channels are found in the membrane of most cells and they allow positively or negatively charged ions to pass in and out. A theory attempting to explain how the movement of ions produced the nerve impulse had first been proposed by the German Julius Bernstein in 1902. But it was the work of Hodgkin and Huxley that ignited the field in 1952, when they provided a mathematical description of how sodium ($Na+$) and potassium ($K+$) ions flowed through the nerve axon membrane. Despite this, the existence of ion channels was far from proven, not least because they are far too tiny to be seen under the most powerful microscope. There was also much speculation concerning the mechanism by which the ion channel opened. Most researchers believed the ion channel was some type of specialised "pore" that only allowed the appropriate ion through at precisely the right moment. But how it managed this feat was unclear. Yet, there were other possibilities including one where the ions were carried through the membrane on some type of transporter system. In fact, this was the option that Hodgkin and Huxley had considered most likely. Unfortunately, for neurophysiologists at the time, there seemed no way the ion channel question could ever be answered or resolved.

In 1970, Bernard Katz at University College, London, along with Ricardo Miledi, made a breakthrough. They noticed, as had others, that if they filled a micropipette with acetylcholine and bought it close to the neuromuscular junction, a recording electrode placed inside the muscle cell would show slightly increased electrical activity. Although it appeared as little more than "noise", Katz and Miledi reasoned that it must be the result of small packets of acetylcholine diffusing out of the micropipette and binding with the receptors on the muscle's surface. From this insight, Katz and Miledi were able to apply statistical methods to analyse the intracellular activity – from which they could estimate the average time an individual channel was open (a few milliseconds) and the amount of ionic current that passed through it during this time. It was a clever technique, although much about the functioning of ion channels still had to be learned.

Sakmann had bought experience of using the micropipette technique back from London after working with Katz – and the two men now set about trying to improve it by obtaining more accurate measurements. One of their first innovations was to use denervated supersensitive frog muscle fibres, where the nerve that innervates the muscle is lesioned, which leads to the receptors becoming more sensitive to acetylcholine. But their real key to success came when they found a way of isolating a tiny part of muscle membrane that was small enough to contain an individual ion channel. To achieve this, they took advantage of a newly developed micropipette with an extremely small opening of between 0.5 and 1 mm (one-thousandth of a millimetre). By using its suction force, Neher and Sakmann were able to tear away a tiny piece of membrane. They called this the *patch-clamp method*. Neher and Sakmann realised that any ions flowing through this membrane passed through the channel. Moreover, the current, no matter how small, could be measured by an electrode placed in the pipette, attached to a very sensitive amplifier. Unfortunately, when Neher and Sakmann attempted to do this, they were unable to establish a complete electric seal between pipette and membrane. Thus, the current flowing through the pipette was not identical to that flowing through the membrane. Despite this restriction, Neher and Sakmann eventually managed to produce vague square-like "blips" on their oscilloscope, which provided a measure of the current passing through a single ion channel. These blips also appeared to show that the ion channels opened in an all-or-none manner. It was an encouraging start considering the limitations of their technique – and they introduced their patch-clamp method to the scientific community in 1976.

To their great frustration, Neher and Sakmann's attempts to perfect the seal were continually thwarted until one day in 1980 when they noticed its effectiveness was dramatically increased if a gentle suction was applied to the inside of the pipette. By doing this, they increased the electrical resistance between the wall of the micropipette and membrane over a thousand times. Suddenly, their oscilloscope signal that had looked like small blips atop a sea of static now became clearly square-like. And it was sufficient to characterise individual ion channels with a precise amplitude and time resolution. The results now allowed Neher and Sakmann to observe an ion channel as it opened and closed in real time. In fact, it enabled Neher and Sakmann to detect currents of approximately 1–2 picroamperes (pA), which is equivalent to some 0.000000000001 amps. They also observed a single ion channel activated by acetylcholine opening and shutting repeatedly – with its switching time being less than 10 picoseconds (a picosecond is one millionth of one millionth of a second). Furthermore, through careful analysis, Neher and Sakmann were even able to determine that the main sodium current in cultured muscle cells had a mean amplitude of 1.6 pA, which only lasted for a millisecond. From this, they could calculate that approximately 10,000 sodium ions flowed through a channel activated by acetylcholine in any one millisecond.

The patch-clamp method initiated a revolution in our understanding of plasma membrane ion channels and transport mechanisms in a wide range of cells, although it's major use has been to investigate neurons. For example, it has shown that two types of ion channel exist in neurons: those opened by a neurotransmitter binding to a receptor (ligand-gated) and those opened by a change in the internal voltage of the cell (voltage-gated). Both these groups are now known to have several different types of channel for a variety of ions. Consequently, channels are now described in terms of their ionic selectivity (e.g. Na+, K+) and mechanism (voltage- or ligand-gated). This has also helped throw new light on old research findings. For example, patch-clamping has confirmed the predictions of Hodgkin and Huxley by showing that the squid axon contains two types of channels that are permeable to Na+ and K+ ions. Patch-clamping has also revealed that these are voltage-gated, with the sodium channel opening before the potassium one, although the latter remains open for longer. But patch-clamping has done more than answer old questions. It has opened up an entire scientific realm that is inaccessible in any other way. Before its invention, ionic currents could only be estimated using relatively large cells, which could be penetrated with two microelectrodes. Now recordings can be made from almost any type of cell in the body. Moreover, by applying different neurotransmitters, or drugs, to the pipette, the electrical characteristics of cells to chemical stimulation can be established. Given the ubiquity and importance of ion channels in the nervous system, it is not surprising that the invention of the patch-clamp has led to a plethora of new neuroscientific knowledge concerning the nervous system and brain.

It should also be noted that the patch-clamp method has provided many new insights into the molecular basis of disease. Indeed, a number of neurological diseases are caused or affected by the defective functioning of ion channels, which include multiple sclerosis, epilepsy, involuntary motor movement (ataxia) and deafness. These and many other diseases, along with chronic pain syndromes, have become subject to drug therapy that targets the activity of ion channels. One recent estimate has put the sales volume of drugs that act on ion channels at around $6 billion.

In 1991, Neher and Sakmann were awarded the Nobel Prize for "their discoveries concerning the function of single ion channels in cells". By then, the two men had ended their research collaboration that had lasted for 16 years, although both were still actively engaged in extending the uses of their technique. Most notably, Sakmann had turned to using patch-clamping with recombinant DNA technology that allowed different types of ion channels to be expressed in cultured cells. He also developed a method to make patch-pipette recordings of synaptic currents of visually identified neuronal somata and dendrites in brain slices. In 1998, Sakmann moved to the Max Planck Institute in Heidelberg to find the Department of Cell Physiology. He also took up a post as scientific director of the Max Planck Institute in Florida, where he is involved in creating a 3-D map of the brain using fluorescent markers and brain imaging. Neher remained in Germany, becoming director of the Max Planck Institute for Biophysical Chemistry, as well as professor at the University of Göttingen. He is also co-chair of the Bernstein Center for Computational Neuroscience. After his work with Sakmann, Neher adapted the patch-clamping technique to study neurotransmitter and hormone secretion, focusing especially on the molecular events that drive the vesicles to fuse with the membrane. In 2003, Neher was also one of twenty-two Nobel Laureates who signed the Humanist Manifesto supporting a humanist orientation for a better world. There can be little doubt that Sakmann and Neher's invention of patch-clamping has contributed greatly to this cause.

Note

1 Norbert Wiener, who is credited with being the founder of cybernetics in 1948, defined it as "the scientific study of control and communication in the animal and the machine". It can perhaps be more loosely defined as the science of self-automated control.

Sakmann and Neher's major writings

Neher, E. and Sakmann, B. (1976) Noise analysis of drug-induced voltage clamp currents in denervated frog muscle fibres. *Journal of Physiology*, 258, 705–729.

Neher, E. and Sakmann, B. (1976) Single-channel currents recorded from membrane of denervated frog muscle fibres. *Nature*, 260, 779–802.

Neher, E. (1991) Ion channels for communication between and within cells. In: N. Ringertz (ed.), *Nobel Lectures, Physiology or Medicine 1991–1995*. Singapore: World Scientific Publishing (1997).

Sakmann, B., Edwards, F.A., Konnerth, A. and Takahashi, T. (1989) Patch clamp techniques used for studying synaptic transmission in slices of mammalian brain. *Quarterly Journal of Experimental Physiology*, 74, 1107–1118.

Sakmann, B. and Neher, E. (1994) Patch clamp techniques for studying excitable membranes. *Annual Review of Physiology*, 46, 455–472.

Sakmann, B. (1991) Elementary steps in synaptic transmission revealed by currents through single ion channels. In: N Ringertz (ed.), *Nobel Lectures, Physiology or Medicine 1991–1995*. Singapore: World Scientific Publishing. (1997).

Sakmann, B. and Neher, E. (2010) *Single Channel Recording*. Springer: New York.

Further reading

Aidley, D.J. and Stanfield, P.R. (2008) *Ion Channels: Molecules in Action*. Cambridge, UK: Cambridge University Press.

Liem, L.K., Simard, J.M., Song, Y. and Tewari, K. (1995) The patch clamp technique. *Neurosurgery*, 36, 382–392.

McComas, A.J. (2011) *Galvani's Spark: The Story of the Nerve Impulse*. Oxford: Oxford University Press.

Sigworth, F.J. (1986) The patch clamp is more useful than anyone had expected. *Federation Proceedings*, 45, 2673–2677.

Tao, C., Zhang, G., Xiong, Y. and Zhou, Y. (2015) Functional dissection of synaptic circuits: *In vivo* patch-clamp recording in neuroscience. *Frontiers in Neural Circuits*, 9, 23. DOI:10.3389/fncir.2015.00023.

Chapter 30

David Coutnay Marr (1945–1980)

David Marr was an English theoretical neuroscientist, who attempted to model how the brain works by using a combination of hard biological fact and computational analysis – work that was to have its biggest impact on explaining vision, although he also made important contributions to the understanding of the cerebellum, hippocampus and neocortex.

David Marr was born in Woodford, a suburban middle-class town of East London, in 1945. Little appears to have been written about his early life, although it is clear that he enjoyed and excelled in music, being an accomplished clarinettist with a love for Berlioz, Schubert and Mozart. He also sometimes played in Britain's National Youth Orchestra. From 1958 to 1963, Marr was educated at Rugby, an all-boys English public boarding school that had been immortalised in the 1857 novel *Tom Brown's School Days*. Here, Marr received a broad, rigorous and challenging education, developing an interest in mathematics and physics while least liking sports. Marr also learned to speak Russian, and by the end of his schooling, he was talking enthusiastically to friends about the possibility of developing a mathematical theory of brain function, after reading Grey Walter's book *The Living Brain*. By then, he had also obtained his pilot's licence. At the age of 18, Marr won a scholarship that allowed him to enter Trinity College, Cambridge, where he was described by one tutor as being serious, with a love of exactness, yet endearing with his bubbly laughter. Marr took just 2 years to obtain a BA degree in mathematics with first-class honours in 1965 and would follow this with an MA degree in 1966. He was then awarded the Coutts Trotter Scholarship that enabled him to take courses in neurological subjects at

Cambridge for a year without having to take any examinations. The main reason for doing this extra study was to prepare him more fully towards a doctorate in theoretical neuroscience, which he did under the supervision of Giles Brindley,[1] who was known for his work on the retina and colour vision. The choice of Brindley may also have been influenced by his proficiency as a classical bassoonist, with the two often playing music together, and Beethoven being a favourite. Cambridge also proved an inspirational environment for a young bidding neuroscientist as Marr benefited from lectures and interactions with a number of luminaries including Alan Hodgkin, William Ruston and Horace Barlow – the latter introducing him to the work of Donald Hebb.

Marr submitted his dissertation in July 1968. Although he had not undertaken any experimental research, his theorising led to the publication of three journal papers – the first appearing in 1969, which described a theoretical model of the *cerebellum* based on anatomical and physiological data (the other two were on the *hippocampus* and *neocortex*). Marr had provided a highly original and pioneering approach to understanding how the cerebellum computed information, and it resulted in him winning a Prize Research Fellowship – a remarkable feat since his PHD had taken barely a year to complete. In 1971, Marr accepted a research appointment at the Medical Research Council laboratory of molecular biology (also in Cambridge) under Sydney Bremer and Francis Crick. Here, he gave several seminars based on his papers and worked mainly on a theoretical understanding of the neocortex and hippocampus. Marr continued this work at Kings College, Cambridge, from 1972 to 1976, although from 1973 onwards he would increasingly spend time in the artificial intelligence laboratory at the Massachusetts Institute of Technology. Here, Marr came into contact with a number of artificial intelligence (AI) experts including Marvin Minsky and Seymour Papert. This would lead to a faculty appointment in the Department of Brain and Computer Sciences at MIT in 1977 and full professorship in 1980.

Marr's doctoral work on the cerebellum, published in the *Journal of Physiology* in 1969, provides a good example of his computational approach to understanding the brain's neural function. The cerebellum is an evolutionary old structure, which has become especially intricate in humans, making up only about 10% of our brains mass, yet containing some 50% of its neurons. Traditionally, it has been recognised as having an important role in the regulation of motor control – as can be seen by the effects of cerebellar damage in humans, which produces a lack of fluidity in voluntary movement along with postural and balance difficulties. The cerebellum's neuroanatomy was first fully illustrated in 1888 by Ramón y Cajal, who revealed several types of neuron, including large Purkinje cells (first described by Purkinje in 1837) whose axons provided the sole cerebellar output to the rest of the brain. During the 1960s, John Eccles and others determined the excitatory and inhibitory nature of each cell type. The Purkinje cells were shown to be inhibitory, whereas the incoming parallel fibres and climbing fibres (which both project to the Purkinje cells) were excitatory. In fact, the climbing fibres, which intertwine themselves around the dendrites of the Purkinje cells, make literally hundreds of synaptic connections and constitute one of the most powerful excitatory connections in the entire CNS.

It was against this backdrop that Marr began his theorising. In 1964, Giles Brindley had pointed out that the convergence of excitatory parallel and climbing fibres onto Purkinje cells implied the existence of Hebbian synapses.[2] More specifically, he reasoned: if the activity of the parallel and climbing fibres was synchronised, then the synapses between the parallel fibres and the Purkinje cells would become strengthened – fulfilling Hebb's criteria for the occurrence of neural plasticity (i.e. learning). This also implicated the cerebellum in the learning of new motor skills. Marr extended this idea by proposing that each climbing fibre conveyed an instruction from the cerebral cortex to the Purkinje cell regarding an *elemental movement* (such as a fine digit movement or limb action). At the same time, the parallel fibre input to the Purkinje cell was seen as providing information about the situational context in which the climbing fibre fired. This meant, according to Marr, during the learning of a new skill, each Purkinje cell began to recognise the contexts in which the action was appropriate. And, ultimately, this led to a situation where once a movement had been learned, the context alone could be enough to fire the Purkinje cell and its elemental action. Thus, the movement became automated. It was also a theory that placed the memory of skilled motor behaviour, at least in part, in the synapses located between the parallel fibres and the Purkinje cells.

Marr's paper is widely recognised as a milestone in the history of the cerebellum, which not only integrated information from neuroanatomy and physiology but also produced several predictions that could be experimentally tested and adopted in computer modelling. Needless to say, it generated an enormous amount of research. Perhaps the most important prediction of the theory was that motor learning should be accompanied by facilitatory changes in the synaptic strength between the parallel fibres and the Purkinje cells – especially if there is synchronised climbing fibre activity. This theory has been tested by Masao Ito and his colleagues at the University of Tokyo, who recorded from the cerebellum in animals made to undergo a procedure whereby the eyes remain fixed on an object while the head is turning (known as the vestibulo-ocular reflex). Their work has shown that when a Purkinje cell receives synchronised input from the climbing and parallel fibres, it becomes persistently depressed (or inhibited). Although this contradicts Marr, who thought such synchronisation should lead to Purkinje excitation, it does not refute his general idea of cerebellar learning. Nor does it alter the fundamental role of Purkinje cells in his theory.

Marr also constructed a theory of computational information processing to explain the role of the hippocampus (or what he called "archicortex") in memory. According to Marr, the process of memory storage first begins when events in the outside world are represented by patterns of neural activity in associative areas of the neocortex. Following this, the hippocampus acts as a temporary memory store, laying down in its modifiable synapses a very simplified representation of the input to be learned. In turn, this provides a template by which the neocortex can greatly elaborate the memory and make a longer term representation. Marr also recognised that the hippocampus was involved in memory recall. This could occur, he reasoned, when a partial activation of the neocortical memory of an event led to the complete re-activation of a simple representation in the hippocampus – which in turn was able to trigger the entire neocortical store of information regarding that memory. These were innovative ideas for the early 1970s, which set the stage for several decades of subsequent research.

But Marr's greatest contribution to computational neuroscience has been in vision – which according to Richard Gregory represents the most important contribution of anyone since the work of von Helmholtz. For Marr, vision is "exactly and precisely an information-processing problem" and to solve it one has to discover the nature of the computations performed by the system. In his posthumous book *Vision*, Marr begins by arguing that this problem (or indeed any form of machine learning) has to be tackled at three different levels – computational, algorithmic and implementation. The computational stage has to recognise the overall goal and logic of the strategy that needs to be carried out and is the foundation on which the theory will have to be based. The second algorithmic stage attempts to address the problem of computation. In particular, it asks: what is the mathematical representation of the input and output? And what is the algorithm (i.e. the set of rules) necessary to undertake this computational transformation. Finally, the theorist has to determine how the representation and algorithm can be realised in a physical system. To solve this final level, one needs to study the brain and understand the neuroanatomy and neurophysiology of the processes underlying perception, cognition and action.

The crux of Marr's approach to vision, however, lies with nature of the computations the visual system performs in order to transform a retinal image into a rich visual experience. Marr saw this as a process with four distinct stages. The first occurs (1) when the retina turns light information into neural activity that is relayed to the brain by its ganglion cells. At this point, Marr argues that the image can be likened to an array of black-and-white pixels, which he calls "grayscale" not unlike a picture on a black-and-white television screen. This is followed by (2) the creation of the primal sketch (or outline) that provides a two-dimensional description of the visual input, including information about edges, contours and blobs. According to Marr, this sketching process is undertaken by the simple cells of the visual cortex, although he did not believe that they detected image components as feature detection theory holds, but rather they applied weighting functions to different pixels. After this analysis, other global features are added to the sketch, (3) including shading, texture, motion and binocular disparity. Marr called this third stage the full primal sketch. It should also be noted that Marr's modular approach to perception also allows other visual information to be processed simultaneously while the full primal sketch is being created – and much of this has to do with establishing colour and depth relations. Finally, in the final stage of visual processing, (4) 3-D representation occurs, whereby the object can be recognised from any angle (i.e. independent of the viewpoint of the observer). Roughly speaking, therefore, these four stages can be described as "image-based", "surface-based", "object-based" and "category-based".

Although Marr was not the first to advocate a computational approach to understanding the brain, there is little doubt that his remarkable insight and capacity to unify ideas from neurophysiology, anatomy, mathematics and computing were almost single-handedly responsible for creating a new intellectual landscape that continues to be explored today. His work also fits in with the zeitgeist of the times that is dominated by digital technology and the computer. But not all agree that the computer provides a useful metaphor for the brain. In fact, a typical digital computer uses serial processing, in which only one computation at a time can be performed – but at very high speed. A human brain, in contrast, processes information much more slowly and is composed of a massively complex distributed system of neurons, where a large number of relatively simple computations appear to occur in parallel. It still remains unclear how a neural network system as occurs with the brain can perform many of the algorithmic functions envisaged by Marr. Nonetheless, maybe, one day, these two approaches will become unified in a computational theory of vision.

Tragically, in 1978, Marr was diagnosed with acute leukaemia and he died 2 years later at the age of just 35 in 1980. Nearly 40 years on, his influence is still strongly felt in neuroscience, and there are several awards and prizes named in his honour. These include the Marr Prize, which is one of the most prestigious awards in computer vision, and the David Marr Medal, awarded every 2 years for research into applied vision. His book *Vision* still remains a classic in the field. During his short life, Marr's star burned brightly, and his legacy will surely live on as the founder of theoretical neurobiology and one of the most influential neuroscientists of the twentieth century, who grappled with some of its deepest problems.

Notes

1 Brindley is also infamously known for his unusual behaviour at the 1983 Las Vegas meeting of the American Urological Association, where he removed his pants to show the audience his chemically induced erection. He had injected into his penis a solution of phenoxybenzamine before the presentation.

2 As Hebb wrote in *The Organization of Behavior* (1949): "When an axon of cell A is near enough to excite cell B and repeatedly or persistently takes part in firing it, some growth process or metabolic change, takes place in one or both cells such that A's efficiency . . . is increased". This can be simplified as "neurons that fire together wire together".

Marr's major writings

Marr, D. (1969) A theory of the cerebellar cortex. *Journal of Physiology*, 202, 437–40.

Marr, D. (1970) A theory for cerebral neocortex. *Proceedings of the Royal Society of London, B*, 176, 161–234.

Marr, D. (1971) Simple memory: A theory for archicortex. *Philosophical Transactions of the Royal Society of London, B*, 262, 23–81.

Marr, D. and Hildreth, E. (1980) Theory of edge detection. *Proceedings of the Royal Society of London, B*, 207, 187–217.

Marr, D. (1982) Vision: *A Computational Investigation into the Human Representation and Processing of Visual Information*. New York: W.H. Freeman.

Further reading

Glennerster, A. (2007) Marr's vision: Twenty-five years on. *Current Biology*, 17, pR397–pR399.

Gordon, I.E. (2004) *Theories of Visual Perception*. Hove, East Sussex: Psychology Press.

Vaina, L.M. (1991) *From the Retina to the Neocortex: Selected papers of David Marr*. Berlin: Birkhüuser.

Vaina, L.M. (2006) Marr, David. *Encyclopedia of Cognitive Sciences*. Wiley Online Library, published online. DOI:10.1002/0470018860.s00456.

Vaina, L.M. and Passingham, R.E. (2016) *Computational Theories and their Implementation in the Brain: The Legacy of David Marr*. Oxford: Oxford University Press.

Willshaw, D.J., Dayan, P. and Morris, R.G.M. (2015) Memory, modelling and Marr: A commentary on Marr (1971) 'Simple memory': A theory of archicortex. *Philosophical Transactions of the Royal Society of London, B*. DOI:10.1098/rstb.2014.0383.

Chapter 31

Mortimer Mishkin (b. 1926) and Leslie Ungerleider (b. 1946)

Mishkin and Ungerleider are best known for demonstrating the existence of two distinct visual systems in the primate brain, one coursing through the temporal cortex and the other through the parietal cortex, which are involved in object recognition and spatial awareness, respectively.

Mortimer Mishkin was born in Fitchburg, Massachusetts, to parents he once described as strongly supportive. One of three children, Mishkin attended high and grammar school in Fitchburg, but left a semester before graduation in 1944 to enter the Navy Officer Training Program. A keen musician able to play the saxophone and clarinet, Mishkin had also read Sigmund Freud with interest at school, although his ambition was to become a pilot in the V5 program. However, by 1944, the US Navy had enough pilots to finish the war, and Mishkin joined the Supply Corps – a position he thought might allow him to take some courses in psychology. Indeed, Mishkin would complete a bachelor's degree in business administration at Dartmouth College, New Hampshire, in 1946 which was a necessary prerequisite for Supply Corp training. But the psychology component of his degree was not quite as Mishkin had envisaged, for he was only able to take two courses: mental hygiene and international relations. After graduation, Mishkin was sent to Japan where he was employed in the Tokyo dockyards checking the cargo of oil tankers arriving from the Arabian Sea. In 1947, Mishkin returned to the United States with the intent of learning social psychology. Because he had so few relevant qualifications, Mishkin resorted to getting two letters of recommendation from Dartmouth College. Remarkably, these were sufficient to get Mishkin enrolled into Montreal's ivy league McGill University on a master's degree under Donald Hebb. Here, he built a tachisoscopic (an apparatus that presents visual stimuli at fast speeds) and examined the left versus right visual field recognition of words. After his masters in 1949, Mishkin began working with Haldor E. Rosvold, who was assessing the effects of frontal lobotomy on IQ in schizophrenic patients. But Mishkin's main aim was to undertake brain research, and when Rosvold left for Yale, he was able to follow him, despite being registered as a PhD student at McGill. Mishkin would soon come under the influence of neurosurgeon and primate experimenter Karl Pribram, who had just arrived from working with Karl Lashley in Orange Park, Florida. In fact, both Rosvold and Pribram supervised Mishkin's PhD that examined the effects of temporal lobe lesions in monkeys on visual perception and learning. Mishkin obtained his PhD (from McGill University) in 1951.

Soon after obtaining his PhD, Mishkin and Pribram moved to the curiously named Institute of the Living in Hartford, Connecticut – a private residential psychiatric facility for wealthy patients. The reason for Mishkin's move, however, was that the Institute had decided to set up a research laboratory to examine the behaviour of brain lesioned primates. Although some of Mishkin's work in Hartford examined the consequences of frontal lobe lesions on tasks requiring delayed responding (a task that such animals find difficult), his most interesting work was on the temporal lobes and their involvement in visual processing. Mishkin had begun this line of enquiry to follow-up work by Heinrich Klüver and Paul Bucy undertaken in the late 1930s, who had found that bilateral lesions[1] of the anterior temporal lobes produced an unusual constellation of symptoms in the monkey – including a form of visual agnosia, whereby the animal seemed unable to recognise familiar objects despite having no obvious deficit in seeing. Mishkin not only replicated these effects but also located the visual deficit more precisely to the anterior middle and inferior temporal gyri – a region that become known as the IT cortex.

Further research examining the effects of bilateral IT cortex lesions showed that they impaired the monkey on a wide range of visual discrimination tasks, including the ability to distinguish different shades of gray or determining the sizes of shapes. And as the complexity of the task increased, so did the lesion deficit. From such work, it became apparent that the IT cortex was necessary for normal pattern recognition – a somewhat unexpected finding because at the time it was believed that the only brain area to be involved in vision was the primary visual (striate) cortex located within the occipital lobes. Indeed, one of the doyens of neuroscience, Karl Lashley, had shown in 1948 that large lesions to the association areas surrounding the striate cortex produced few, if any, detrimental effects on visual habits. Thus, for Lashley, the faculty of vision was localised firmly in the striate cortex itself. But the work of Mishkin was showing otherwise and it would be the start of a long quest for him trying to understand how the temporal lobes were involved in the processing of visual information.

In 1955, Mishkin moved to the National Institute of Mental Health where Haldor Rosvold had been appointed to head a neuropsychology department. Here, Mishkin continued his work on the temporal lobe. It was now becoming clear that bilateral IT lesions impaired the ability of monkeys to recognise, learn and remember visual information, although this did not result from any form of blindness or change in visual acuity. While these results were puzzling, they were not unprecedented. In fact, it had been known since Freud that temporal lobe damage in humans could produce visual agnosia where patients could see objects but not recognise them – a deficit that was thought to be more to do with learning and memory than vision. And further evidence for a temporal lobe involvement in vision came in the 1960s from a post-doctoral fellow at the Massachussets Institute of Technology called Charles Gross, who showed that single cells in the IT cortex had large receptive fields that responded exclusively to complex visual stimuli. Indeed, unlike the cells in the striate cortex, which as Hubel and Wiesel had shown responded to spots, slits and edges, those in the IT cortex were sensitive to much more complex visual patterns, including the experimenter's finger, a burning Q-tip or a bottle brush. Later, some single cells in the IT cortex would be shown to selectively fire to the presentation of faces and hands.

One of the problems facing Mishkin at this stage was trying to understand how visual information reached the IT cortex. Although the IT cortex was known to receive input from subcortical visual structures such as the pulvinar and superior colliculus, lesions to these areas did produce the same impairments on visual discrimination tasks. Hence, a more obvious explanation was that the IT cortex received visual input from the striate (visual) cortex. But Lashley had seemingly ruled this out. However, in 1965, the neuroanatomist Hans Kuypers showed that a multisynaptic pathway did exist – which extended from the association striate areas (adjacent to the primary visual cortex) to anterior regions of the temporal lobe. A close examination of this pathway also revealed that it was particularly extensive and difficult to lesion fully, with parts of the pathway buried away in sulci or intertwined with other parts of the visual system such as the optic radiations. Consequently, Lashley's attempted lesioning of this pathway had been far from complete, and it could explain why he had found so little visual impairment.

Nonetheless, it remained to be proven that damage to this pathway was the prime cause of the temporal lobe visual deficit, and Mishkin would do this in an ingenious study using serial lesioning. To begin, he unilaterally ablated the IT cortex. As expected, this produced no visual deficit since it was well established that a bilateral lesion was needed. He then removed the striate cortex in the opposite hemisphere. Again, this produced no deficit on a visual discrimination task (because the remaining striate cortex could still send visual input to the opposite-sided IT cortex). But in the final stage, Mishkin severed the corpus callosum, and this time observed a marked visual impairment. It was not only proof of a complex functional pathway between the striate and IT cortex but also that it was involved in higher order visual processing. A second major advance in understanding the IT cortex around this time was Mishkin's discovery, made with Japanese researcher Eichii Iwai, that the IT cortex could be divided into two regions: an anterior region (TE) and a more posterior area (TEO). The former area was shown to play a more important role in mnemonic aspects such as recognising a recently seen object, whereas the latter area had a more important perceptual function with marked deficits shown on single pattern discrimination tasks.

In 1975, a postdoctoral student called Leslie Ungerleider moved to NIMH where she joined Mishkin's research group. Ungerleider had grown up in New York and gone to Binghamton University, graduating with a BA in experimental psychology in 1966. Although she had considered studying medicine, brain science was the simpler option, and Ungerleider took her PhD at New York University, investigating the behavioural functions of the hypothalamus with electrical stimulation techniques. After a 2-year hiatus in a faculty position at Oklahoma City University – a destination where her husband was forced to complete his medical training – Ungerleider obtained a fellowship in 1972 to work with Karl Pribram at Stanford, where she examined the effects of cortical lesions on the higher order perceptual abilities of monkeys. It was through this relationship she first met Mishkin, and on joining his laboratory, Ungerleider devoted herself to using anatomical tracing techniques in the monkey brain to map out the increasing number of areas that were then being found to comprise the visual association cortex and their interconnections with the temporal lobes.

In 1982, Ungerleider and Mishkin published a paper that became a classic in the field of visual neuroscience. Based on extensive electrophysiological, anatomical and behavioural evidence, including much of their own research, they proposed the existence of two anatomically distinct cortical visual systems in the rhesus monkey's brain. As the work of Mishkin had already shown, one of these pathways extended ventrally from the visual cortex into the temporal lobes and was responsible for the recognition of objects. It soon colloquially became known as the "what is it?" pathway. The other system extended dorsally from the visual cortex through the parietal lobes and was deemed to be specialised in locating objects in space. It soon became known as the "where is it?" pathway.

Strong support for this dissociation came from behavioural studies. For example, in the landmark task, monkeys were presented with two food wells. The animal was then rewarded when it chose the well that was closer to a striped cylinder (i.e. the landmark). This task, therefore, assessed the monkey's ability to recognise the baited well based on its spatial position to the landmark. Monkeys with posterior parietal lobe lesions found this task difficult, unlike those with IT cortex damage. However, things were reversed where the food wells were marked with a place card (either a positive sign or the outline of a square) and the monkey required to differentiate between them. Here, lesions to the IT cortex disrupted the visual discrimination, in contrast to parietal lesions that produced little behavioural deficit.

The theory of separate systems in the brain for visual object recognition and spatial location was a major breakthrough, which helped synthesise a large amount of data into a coherent whole. It was also applied to humans, most notably through the work of Melvyn Goodale and David Milner, who changed the theory by proposing that the main function of the parietal pathway is to visually guide motor actions. For example, if someone throws you a ball, you are likely to try catching it – although you don't plan the catch or think about it. Because of this, Goodale and Milner believe that the parietal pathway is largely involved in automatic and unconscious behaviour. This is in contrast to the temporal lobe (ventral) pathway that is concerned with the conscious recognition of objects in the visual world. Goodale and Milner have also backed up their claims by reporting the case of a patient known as DF, who suffered damage to the ventral stream from carbon monoxide poisoning. Despite her inability to recognise the size, shape and orientation of visual objects, DF was able to accurately guide her hand and fingers to manipulate the very same objects.

Over the long course of his research career, which now approaches 70 years, Mishkin has published hundreds of research papers and book articles, utilising a wide range of interdisciplinary techniques, aimed at elucidating the function of many brain areas. Although best known for his work on the temporal lobes, he has made important contributions to an understanding of the frontal lobes, striatum, visual cortex and several subcortical regions. And in addition to his work on vision, he has made equally important contributions to the neurophysiological basis of memory. At the time of writing, Mishkin was chief of cognitive neuroscience at NIMH, and among his many awards, in 2010, he was presented with the National Medal of Science by Barack Obama. Like Mishkin, Ungerleider remains at the NIMH. She established her own laboratory there in 1995 and now heads the Brain and Cognition Laboratory. She is presently engaged in understanding human attention and perception using brain scanning techniques such as PET and fMRI. In 2001, she was the recipient of the Women in Neuroscience Lifetime Achievement Award, and in 2008, she became an NIH distinguished investigator. One of the few women to head an esteemed department of brain science, Ungerleider will hopefully be a role model for many other aspiring women to reach the top in neuroscience.

Note

1 It is important to emphasise that visual deficits are only obtained after bilateral lesions of the temporal lobes. Unilateral lesions do not produce any noticeable impairment. The reasons for this are complex though best understood by realising that the striate cortex has unilateral projections to the IT cortex, as well as bilateral ones that cross in the corpus callosum.

Mishkin and Ungeleider's major writings

Kuypers, H.G.J.M., Szwarcbart, M.K., Mishkin, M. and Rosvold, H.E. (1965) Occipitotemporal corticocortical connections in the rhesus monkey. *Experimental Neurology*, 11, 245–262.

Mishkin, M. and Ungerleider, L.G. (1982) Contribution of striate inputs to the visuospatial functions of parieto-preoccipital cortex in monkeys. *Behavioral Brain Research*, 6, 57–77.

Mishkin, M., Ungerleider, L.G. and Macko, K.A. (1983) Object vision and spatial vision: Two cortical pathways. *Trends in Neurosciences*, 6, 414–417.

Mishkin, M. (1990) Vision, memory, and the temporal lobe: Summary and perspective. In: E. Iwai and M. Mishkin (eds), *Vision, Memory and the Temporal Lobe*. Amsterdam: Elsevier.

Ungerleider, L.G. and Mishkin, M. (1982) Two cortical visual systems. In: D.J. Ingle, R.J.W. Mansfield, R.J.W. and Goodale, M.A. (eds), *The Analysis of Visual Behavior*. Cambridge, MA: MIT Press.

Ungerleider, L.G., Galkin, T.W. and Mishkin, M. (1983) Visuotopic organization of projections from striate cortex to inferior and lateral pulvinar in rhesus monkey. *Journal of Comparative Neurology*, 217, 137–157.

Further reading

Findall, J.W. and Gonzalez, C.L.R. (2017) Revisiting Ungerleider and Mishkin: Two cortical visual systems. In: B. Kolb and I. Whishaw (eds), *Brain and Behavior: Revisiting the Classic Studies*. Los Angeles: Sage.

Goodale, M. and Miller, D. (2004) *Sight Unseen*. Oxford: Oxford University Press.

Gross, C.G. (1998) *Brain, Vision, Memory*. Cambridge, MA: MIT Press.

Gross, C.R. (1994) How inferior temporal cortex became a visual area. *Cerebral Cortex*, 5, 455–469.

Schenk, T. and Mcintosh, R.D. (2010) Do we have independent visual streams for perception and action? *Cognitive Neuroscience*, 1, 52–62.

Chapter 32

Timothy Bliss (b. 1940) and Terje Lømo (b. 1935)

Bliss and Lømo are best known for their discovery of long-term potentiation – a phenomenon whereby brief high-frequency impulses at certain synapses in the brain produce long-lasting increases in the efficacy of neurotransmission – a property believed to underlie the neural basis of learning and memory.

Terji Lømo was born into a middle-class family (his father was a dentist) in Ålesund, a small town of three islands that lie against a dramatic backdrop of fiords and steep mountains in western Norway. Growing up in an occupied country during World War II, Lømo's early life seems to have been uneventful and he went to Ålesund's only high school where he recalls studying mainly Norwegian, mathematics and physics. By his own admission, Lømo was not an enthusiastic student and often suffered from "low spirits" and depression. Maybe because of this, he did not obtain particularly high marks at school. Despite this, in his early teens, he cycled around England one summer, and in 1951, he won a scholarship to study for a year at a high school in Fort Wayne, Indiana. By his teens, Lømo had decided to become a doctor, and in 1954, at the age of 19, he enrolled in medical school at the University of Oslo – the only one in the country. Lømo has written that he barely qualified for medical studies and only did so because he was born in the Great Depression when there were few births in Norway and consequently the number of applicants was correspondingly low. It perhaps did not help his cause that he also took an interest in left-wing politics. Nonetheless, at medical school, Lømo came under the tutelage of Alf Brodal, who was a renowned neuroanatomist, from whom he developed an interest in the tracing of neural pathways in the brain. In the midst of his studies, however, Lømo took a break from Oslo in 1958 by securing a fellowship to work at the University of Pisa. Employed in the laboratory of sleep researcher Giuseppe Moruzzi, Lømo learned stereotaxic surgery and spent time recoding from single neurons in the rabbit cerebral cortex – work leading to his first scientific paper. In 1959, Lømo returned to Norway and moved to Bergen where he finished the final 2 years of medical school. After obtaining his MD, he served as a general practitioner for a year and then worked as an assistant community doctor in Tromsø, and on the small island of Skjævøy which lies north of the Arctic Circle. Lømo was then forced to complete military service by serving as a doctor in the Norwegian Navy – some of it spent on frigates where he often succumbed to sea-sickness.

In 1964, Lømo was walking the streets of Oslo looking for a job when he bumped into the neurophysiologist Per Andersen. Known for his work on the hippocampus, Andersen was looking for researchers to join his laboratory, and Lømo with his experience of recording from single neurons in Italy was a natural choice. At first, he was utilised as Andersen's research assistant, where among other things he recorded from neurons in the CA1 region of the hippocampus and measured tiny temperature changes along their dendrites. But by the end of 1965, it was time for Lømo to undertake his own PhD. One of the first things that Lømo investigated was a phenomenon that Andersen had first discovered in the hippocampus called "frequency potentiation". To fully understand this phenomenon, one has to know a little about the anatomy of the hippocampus, which contains three layers of cells, not unlike the jam layers in a Swiss roll. The cells that interested Andersen were the granular cells of the hippocampal dentate gyrus, which receive input from the adjacent entorhinal cortex via a bundle of fibres called the perforant path. Andersen had shown that if one stimulates the perforant path with a burst of impulses, it led to increased excitability of the granule cells for several minutes. But as Lømo began to examine the parameters of this effect in more detail, he realised that if the perforant path was stimulated with trains of stimuli for 10 seconds at frequencies of around 10 to 20 Hz,[1] then a single brief stimulus applied later could produce increased excitability in the granule cells, which lasted for several hours. This effect would later become known as *long-term potentiation* (LTP). Lømo was also aware that the hippocampus was implicated in the formation of new memories and frequency potentiation was a good candidate for its neural substrate. Although he presented his findings in a meeting of the Scandinavian Physiological Society in Turku, Finland, in 1966, they did not generate much interest. And perhaps even odder, Lømo turned to other issues in his pursuit of his doctorate that he obtained in 1969.

At around the same time as Lømo was beginning his PhD, an Englishman called Timothy Vivian Pelham Bliss, who had gone to Dean Close boarding school in Cheltenham and whose father had a navel background, was doing his own doctorate on synaptic plasticity at McGill University in Montreal. An ex-physics student, who apparently turned to neurophysiology after hearing a lecture that compared spinal reflexes with the Bell telephone system, Bliss was undertaking single-unit recordings with small slabs of cat cerebral cortex using similar techniques to those of Lømo. However, the neurophysiology of the cortex is much more complex than the hippocampus and Bliss had obtained disappointing results. In 1967, Bliss followed his PhD supervisor Benedict Delisle Burns to London's National Institute of Medical Research at Mill Hill. A year later, Bliss would meet Per Andersen in London. By now, Bliss was considering the possibility of working with the hippocampus and asked Andersen if he could spend some time working in his laboratory. Apparently, during this conversation, Andersen is reported to have said: "Then you should come and talk to Terje Lømo who has found something that will interest you".

In 1968, Bliss began a year research placement in Oslo, working with Andersen and Lømo. This was a period when Lømo was finishing off his own PhD research and could only manage to collaborate with Bliss on 1 day a week. Nevertheless, the two men were able to replicate Lømo's frequency potentiation work – with results exceeding all expectations. For example, they stimulated the perforant path in brain slices and observed the granule cell population spike increase dramatically with each train of impulses, whereas the control pathway on the other side of the hippocampus was unaffected. At first, Bliss was concerned that the effect might be spurious – perhaps because the stimulating electrode tip was lowering the threshold for generating action potentials in the surrounding axons and thus recruiting more fibres to produce the effect. But Bliss and Lømo managed to control this possibility. In short, the results showed that the long-term potential of the hippocampal granule cells was coming about through enhanced synaptic activity – a phenomenon that Donald Hebb had predicted in his book *The Organization of Behavior* in 1949.

Although it was a significant discovery, Bliss and Lømo curiously did not submit their work for publication until 1973 – some 4 years after the experiments. According to Lømo, there were several reasons for the delay. Their results were first presented at a conference – this time in Italy – but again they generated little interest. Soon after, the two men moved to London – Bliss returning to Mill Hill and Lømo taking up a place at University College where he started work with Ricardo Miledi on nerve muscle preparations. Nonetheless, Bliss and Lømo managed to find some time to work on frequency potentiation in anaesthetised rabbits. Unfortunately, they now found the effect hard to replicate. In fact, Lømo would return to Oslo in April 1971 and give up the work. Bliss then formed a collaboration with Tony Gardner-Medwin at University College to start searching for the elusive frequency potentiation again – which they managed to find. This appears to have been the prompt for the publication of the findings from 1969, with two papers appearing together in the *Journal of Physiology*.

There has been speculation as to why frequency potentiation is so variable and capricious to establish. Lømo has written that the most likely explanation is because of differential effects of stress on the animal. In Oslo, a local farmer provided the rabbits directly to the laboratory, whereas in London the animals were handled much more in transit and underwent different experimental procedures. In fact, the two men even jokingly considered importing rabbits from Norway! Lømo also points out that rabbits are particularly sensitive to stress. But perhaps most important, it is now accepted that stress has a particularly potent effect on suppressing LTP in the hippocampus – an effect that occurs *in vivo* and in brain slices taken from animals.

Since the publication of Bliss and Lømo's pioneering work in 1973, thousands of papers have been published on LTP, and compelling evidence shows that it takes place in the neural circuits underlying learning and memory. Yet, it took almost a decade before the work was widely perceived as important. In fact, the first replication of Bliss and Lømo's work was reported in 1975, and soon after LTP was shown in hippocampal slices kept alive in an oxygenated saline bath – an advance enabling the stimulating and recording microelectrodes to be positioned with great precision. This also provided the superior model by which neuroscientists could elucidate the neural basis of LTP. Another key event took place when

Alger and Teyler first used the acronym LTP in 1976 – a change in terminology, which according to Lømo may have won the day, because it sounds better and comes off the tongue more easily. But even so, many neuroscientists did not fully recognise the importance of LTP for understanding brain function until the mid-1980s.

The question of what causes LTP in the hippocampus (although it has also been found in other areas of the brain) has been subject to a large amount of research. The first stage is believed to involve glutamate release from the perforant path that crosses the synapse and binds to receptors located on the dendrites of neurons in the hippocampal dentate gyrus. Here, there are two types of glutamate receptor – the AMPA (α-amino-3-hydroxy-5-methyl-4-isoxazole-propionic acid) and NMDA (N-methyl-D-aspartate). It appears that the initial stimulation of the AMPA receptor causes a moderate depolarisation of the neuron. At this point, however, the NMDA receptor is inactive because of a usual feature of its ion channel, which is blocked by magnesium ions – which stops excitatory calcium ions flowing into the dentate cells. However, when the cell becomes depolarised from AMPA activation, the magnesium ions become displaced, allowing calcium ions to surge into the cell. The result is a delayed but much larger depolarisation of the neuron.

The flow of calcium into the dentate cell is crucial for generating LTP probably because it acts as an intracellular messenger that activates a number of proteins involved in changing cellular function. One such protein is calmodulin, which has a conformation similar to a flower with four petals. When calcium binds to calmodulin, it changes its shape, enabling binding to a group of proteins called protein kinases. One important kinase is the enzyme calcium/calmodulin-dependent kinase II (CaMK II). The activity of this chemical continues long after the calcium levels have returned to normal and mimics to some extent the time course of LTP. Moreover, drugs that inhibit CaMK II are also effective in blocking LTP. The exact mechanism by which CaMK II regulates LTP remains unclear – although one possibility is that it increases the number of glutamate receptors at the synapse in a process called "up-regulation", thereby making the cell more excitable. Indeed, in support of this, calcium increases protein kinase activity that is known to affect chemicals involved in gene transcription.

To make matters more fascinating, there is evidence that LTP is accompanied by increased release of glutamate from the presynaptic perforant path neurons. In other words, as learning takes place, more glutamate is released into the synapse. But, of course, this raises a question: if this occurs, then a retrograde signal must go back to the axon terminals of the perforant path to turn on more "potentiation". Several substances have been implicated in this process, including the gas nitric oxide (NO) that can diffuse through cell membranes. Support for this theory comes from evidence showing that NO is formed by calcium-dependent enzymes which are activated by stimulation of NMDA receptors. Moreover, inhibitors of NO synthase (the enzyme involved in the formation of NO) block the formation of LTP.

The discovery of LTP is a landmark in the history of neuroscience, and one that promises to fully explain the neurophysiological and molecular basis of learning and memory in the brain. However, despite being the prime discoverer of this effect, Lømo would contribute little more to the LTP story after his two 1973 papers with Bliss. By then, he had returned to Norway where he worked on the increased sensitivity of denervated muscle fibres to acetylcholine – an effect he believed occurred through a lack of trophic factors that support the survival of neurons. Although retired, he still has access to his old laboratory in Oslo. Unlike Lømo, Bliss continued his investigation of the cellular properties of LTP at the NIMR where he worked as head of neuroscience from 1968 to 2006. He was also elected to the Royal Society in 1994. Retiring in 2006, Bliss continues to be a visiting professor at the University College London and has strong connections with Xi'an Jiaotong University in China. But maybe the best is yet to come. In a 2016 interview with *The Daily Telegraph*, Bliss maintains it is only a matter of time before memory is fully mastered, and one day in the future it might be possible to visit a "synapse surgeon" who is able to implant a brain with fond memories of spending the night with Marilyn Monroe! If that kind of prospect does not inspire ambitious neuroscience students, then surely nothing will.

Note

1 A hertz is defined as one cycle per second and named after Heinrich Rudolf Hertz who was the first person to prove the existence of electromagnetic waves.

Bliss and Lømo's major writings

Bliss, T. and Lømo, T. (1973) Long-lasting potentiation of synaptic transmission in the dentate area of the anaesthetized rabbit following stimulation of the perforant path. *Journal of Physiology*, 232, 331–356.

Bliss, T. and Gardner-Medwin, A.R. (1973) Long-lasting potentiation of synaptic transmission in the dentate area of the unanaesthetized rabbit following stimulation of the perforant path. *Journal of Physiology*, 232, 357–374.

Bliss, T., Gardner-Medwin, A.R. and Lømo, T. (1973) Synaptic plasticity in the hippocampal formation. In: G.B. Ansell and P.B. Bradley (eds), *Macromolecules and Behavior*. London: MacMillan.

Bliss, T. and Collingridge, G.L. (1993) A synaptic model of memory: Long term potentiation of the hippocampus. *Nature*, 361, 31–39.

Bliss, T., Collingridge, G.L. and Morris, R.G.M. (1993) *Long-term potentiation: Enhancing neuroscience for 30 years*. Oxford: Oxford University Press.

Lømo, T. (2003) The discovery of long term potentiation. *Philosophical Transactions of the Royal Society, London B*, 358, 617–620.

Lømo, T. (2012) Terje Lømo. In: L.R. Squire (ed.), *The History of Neuroscience in Autobiography. Vol 7*. New York: Oxford University Press.

Lømo, T. (2016) Scientific discoveries: What is required for lasting impact. *Annual Review of Physiology*, 78, 1–21.

Further reading

Alger, B.E. and Teyler, T.J. (1976) Long-term and short-term plasticity in the CA1, CA3 and dentate regions of the rat hippocampal slices. *Nature*, 266, 736–737.

Craver, C.F. (2003) The making of a memory mechanism. *Journal of the History of Biology*, 36, 153–195.

Hölscher, C. (ed.) (2001) *Neuronal Mechanisms of Memory Formation: Concepts of long-term potentiation*. Cambridge, UK: Cambridge University Press.

Lisman, J., Lichtman, J.W. and Sanes, J.R. (2003) LTP: Perils and progress. *Nature Reviews: Neuroscience*, 4, 926–929.

Malenka, R.C. and Nicoll, R.A. (1999) Long term potentiation – A decade of progress? *Science*, 285, 1870–1874.

Nicoll, R.A. (2017) A brief history of long-term potentiation. *Neuron*, 93, 281–290.

Chapter 33

Richard Axel (b. 1946) and Linda Buck (b. 1947)

Axel and Buck are famous for their Nobel Prize winning work on olfaction that demonstrated that the nasal cells containing receptors for the detection of smell are encoded by a large family of over 1000 different genes – with each gene specifying a receptor specialised only to detect a few odours.

Richard Axel was born in Brooklyn, just after World War II, to Jewish immigrant parents who had fled from the Nazi invasion of Poland in 1939. Axel's father was a tailor, and although his parents were uneducated, they had a deep respect for learning and encouraged their son to do well at school. Despite this, much of Axel's early education was spent on the streets of Brooklyn playing baseball with a broom handle and using manhole covers as bases. Having no thought of a career in academia, the young teenager took on a number of jobs including delivering false teeth to dentists, laying carpets and working in a delicatessen. Despite this seemingly unambitious start in life, the young Axel must have shown some promise academically, for at the age of 14, he was made to attend Stuyvesant High School in Manhattan, an institution for intellectually gifted boys. At first, Axel was unhappy with this decision as Manhattan was some 10 miles away, and even worse it had the weakest basketball team in the city. But this soon changed as Axel became introduced to art, literature and music, which replaced his obsession with baseball. And when he wasn't attending school, Axel could often be found in the reading room of the New York Public Library on 42nd Street. After leaving high school, Axel attended Columbia College on a scholarship intent on studying literature and poetry. But somewhat unexpectedly, he also took a keen interest in molecular biology after landing himself a job washing glassware in the laboratory of Bernard Weinstein, who was then working on DNA and the genetic code. After gaining knowledge as Weinstein's research student, Axel took the big leap of studying medicine at Johns Hopkins. However, Axel proved to be, in his own words, a terrible medical student. He was not only awkward working with patients (it was in this period he sewed a surgeon's finger to a patient upon suturing an incision) but he did not do much better as an intern in pathology performing autopsies on the deceased (after 1 year, he was told never to practise on dead patients again!). Fortunately, in 1971, Axel was able to return to molecular biology by finding a post back at Columbia where he worked on RNA tumour viruses. After spending a couple of years as a postdoctoral fellowship at the NIH, Axel was then invited back to Columbia as an assistant professor in 1974.

Axel was now in his late twenties and had not set the world alight with his research findings, although things were now about to change. Columbia University was at the forefront of research into recombinant DNA technology and Axel formed a collaboration with Michael Wigler and Saul Silverstein, which allowed him to learn many of its new cutting-edge methods – creating new genetic material, isolating genes and cutting and pasting DNA into new genomes. One benefit of this recombinant engineering was that it could produce many clinically useful proteins from genes inserted into "factory" cells. This new technology would result in a group of five patents (now often referred to as the "Axel patents") that were highly lucrative for Columbia – earning it some $790 million in revenues until they began to expire in 2000.[1] But the financial reward of this new biological engineering was only one aspect. This early work on recombinant DNA was also very exciting since it provided an array of new powerful tools and techniques by which to study the role of genes in behaviour and disease. Not surprisingly, this expertise also attracted a number of researchers to the microbiology department at Columbia University, eager to learn the new technology – one of which was Linda Buck who joined Axel's research group in 1982.

Buck hailed from Seattle, Washington, the second of three children, to a mother with Swedish ancestry and father who was an electrical engineer with a penchant for inventing things. Encouraged to think independently from a young age, and also be ambitious, Buck studied at the University of Washington close to her home. Although she had set out to major in psychology, with a view to becoming a psychotherapist, she increasingly turned to medical biology and graduated with a BSc degree in psychology and microbiology in 1975. In the same year, Buck began her PhD at the University of Texas, examining the functional properties of certain types of white blood cells. In 1980, she moved to Columbia to begin postdoctoral work looking at the specialised proteins found on the surfaces of cells that help the immune system recognise foreign substances. This led to an unexpected discovery when Buck found that some of these proteins also rapidly accumulated inside the cells – a finding that required her to learn more about the molecular mechanisms of cellular biology. To this end, she moved into the laboratory of Richard Axel.

Around this time, Axel was beginning to consider applying the new developments in molecular biology to solving problems in neuroscience – largely as a result of a faculty meeting with Eric Kandel who was also at Columbia. Kandel had identified several behaviours in *Aplysia* (see Chapter 28), which implicated the involvement of genes and protein synthesis, and was keen to elucidate these processes in more detail. It also proved a challenging and interesting problem for Buck to investigate. Her work, published in 1985, would show that an *Aplysia* neuron known as R15, which formed part of its abdominal ganglion, expressed a gene that contributed to producing several different types of neuropeptide. Moreover, these peptides were shown to be involved in several important behaviours of *Aplysia*, including egg-laying and cardiovascular control.

As her *Aplysia* work drew to a close, Buck came across a paper that was to change her life. This concerned the unsolved question of how odours are detected by the nose. The problem was simple: how are humans and other mammals able to detect a huge variety of different smells that were estimated to number 10,000 or more? There was also the second problem of how nearly identical chemicals can generate very different odour perceptions. Buck was immediately hooked by the problem and joined forces with Axel to begin her quest to discover odorant receptors in the nose's epithelium tissue located on the roof of each nasal cavity. The only clue to the puzzle at the time was that the receptors appeared to work through a mechanism involving G-protein-induced increases in the second messenger cAMP. This knowledge led Buck to try several different approaches that would take her more than 5 years to successfully accomplish. But she would hit on a method that entailed the following steps: (1) clone bits of DNA taken from the nose's epithelium tissue, (2) use restriction enzymes to break this DNA into smaller pieces and (3) utilise these pieces as probes (or tags) to identify the genes themselves, which in turn could identify the receptors.

The initial result of her work (Buck and Axel 1991) was a major breakthrough. It showed that in the rat, there was a multigene family that coded for over 100 hundred odorant receptors – all of which were related, yet unique. Moreover, each receptor was a seven transmembrane domain protein that appeared to detect a different smell. In fact, subsequent work showed that humans have around 350 different odorant receptors and mice have about 1000. This indicates that roughly 1 to 5% of the genes in the human genome are devoted to the detection of smells and they form the largest known gene family in our DNA.

In 1991, Buck took up an assistant professorship in the neurobiology department at Harvard Medical School. The next step was to understand how the odorant receptors were organised in the olfactory epithelium. Each olfactory epithelium in the human nose contains specialised bipolar nerve endings of the first cranial nerve that have at their endings some ten to twenty long hairs or cilia. In turn, this sensory nerve projects to the olfactory bulb at the base of each hemisphere where the perception of smell begins to take place. To locate the odorant receptors, Buck, along with graduate student Kerry Ressler, used a technique called *in situ* hybridisation, which used DNA probes to identify specific receptors. This work showed that each olfactory sensory neuron only expressed one type of receptor. From this, Buck was able to estimate that the olfactory epithelium had a total of some five million receptors. Furthermore, it was found that each odorant receptor was widely distributed over the entire olfactory epithelium. These findings were published in 1993, and similar observations in rats were reported in the same year by Axel's group.

Buck's initial results in 1991 had indicated that each odorant receptor was only capable of detecting one type of smell. But the facts of this situation turned out to be more complicated when Buck began working with Takaaki Sato in Japan, who exposed mice olfactory sensory neurons to a series of odorants using calcium imaging to visualise their responses.[2] This research showed that a smell (e.g. orange, nut-like) could activate multiple types of receptor and that a single receptor could be activated by a handful of different odorants at varying intensities. Thus, instead of one odorant for one receptor, it was apparent that the nose was using a combinational code, which was likened to the colours on a patchwork quilt, with different odours being recognised by different combinations of odorant receptors (patches). This theory also helped explain why changing the structure of an odour, even slightly, can significantly alter its perceived smell. The patchwork quilt theory explained this effect by showing that pairs of odorants with very similar chemical structures can produce very different combinations of receptor activation in the cells of the olfactory epithelium.

Following the mapping of the olfactory epithelium, Buck and her colleagues turned their attention to the olfactory bulb – an anatomically complex brain structure that contains a specialised type of neuron called a "mitral cell", which has a triangular shape, named after its resemblance to a bishop's mitre. In the human olfactory bulb, there are about 50,000 mitral cells and each receives input from about 2500 bipolar cell axons originating from the nasal epithelium. Interestingly, the mitral cells are also bundled together in spherical clumps (in essence a dense accumulation of cells, synapses and dendrites) known as olfactory glomeruli – of which there are about 2000 in the human olfactory bulb. Using probes that recognised a single odorant receptor gene, Buck found that each glomerulus receives information from just one kind of olfactory receptor. It was also apparent that each odor presented to the nose activated a different pattern of glomeruli. Put another way, by analysing the different sets of activated glomeruli, one could, in theory, decode the identity of the odor. Again, this work was being paralleled by researchers in Axel's laboratory, who published similar findings for the rats in 1994.

After examining the structure of the olfactory bulb, Buck and her colleagues turned their attention to the higher areas of the brain in receipt of smell information. To do this, she used a novel technique involving transgenic mice that expressed a plant protein in their sensory olfactory epithelium neurons – which was found to travel deep into the brain. By tracing the protein, Buck showed that the neural pathway from the olfactory bulb extended to the pyriform cortex or anterior aspects of the medial temporal lobe – a poorly defined region composed mainly of the amygdala, uncus and parahippocampal gyrus. From here, the olfactory information can reach the mediodorsal thalamus and then prefrontal cortex and parts of the limbic system. These areas are probably responsible for giving meaning and emotional reactions to our olfactory world. Sadly, the role of the pyriform cortex in olfaction has now become somewhat uncertain, with Buck forced to retract three papers on this brain region, which had been co-authored with a postdoctoral student called Zhihua Zou, after being unable to replicate the findings.

In 2004, Buck and Axel shared the US$1.4 million Nobel Prize in physiology or medicine "for their discoveries of odorant receptors and the organisation of the olfactory system". Over the course of some 25 years, Axel and Buck have transformed our understanding of smell and opened up new ways of showing how genes are linked to the transmission of sensory information. Their work has also demonstrated the enormous power by which the techniques of molecular biology can help solve fundamental problems in neuroscience. Both Axel and Buck continue to undertake research. Axel, in addition to his contributions to neurobiology, has also made seminal discoveries in immunology, with his laboratory using recombinant DNA technology in 2004 to discover the molecule CD4, which is a surface protein found on certain types of white blood cells involved in the transmission of HIV. His work also brings great hope of new interventions into the illness. Buck, who returned to Seattle in 2002 to work at the Fred Hutchinson Cancer Research Centre while taking up a professorship at the University of Washington, is now primarily concerned with how pheromones (regulated by a different set of G-proteins than the odorant receptors) are detected in the vomeronasal organ of the nose, as well as looking at the receptor basis of taste detection – a sense closely associated with olfaction.

Notes

1 See Colaianni and Cook-Deegan (2009.)
2 Calcium imaging is a technique that shows the calcium (Ca^{2+}) status of a cell, typically by using fluorescent molecules that respond to calcium ions, which in turn provides an indicator of cellular activity.

Axel and Buck's major writings

Axel, R. (1995) The molecular logic of smell. *Scientific American*, 273, 154–159.

Axel, R. (2004) Scents and sensibility: A molecular logic of olfactory perception. In: T. Frängsmyr (ed.), *Les Prix Nobel. The Nobel Prizes*. Stockholm: Nobel Foundation.

Buck, L.B. and Axel, R. (1991) A novel multigene family may encode odorant receptors: a molecular basis for odor recognition. *Cell*, 65, 175–187.
Buck, L.B. (1996) Information coding in the vertebrate olfactory system. *Annual Review of Neuroscience*, 19, 517–544.
Buck, L.B. (2004) Unravelling the sense of smell. In: T. Frängsmyr (ed.), *Les Prix Nobel. The Nobel Prizes*. Stockholm: Nobel Foundation.
Buck, L.B. (2004) The search for odorant receptors. *Cell*, 116, Suppl 2, S117–S119.
Malnic B., Hirono, J., Sato, T. and Buck, L.B. (1999) Combinational receptor codes for odors. *Cell*, 96, 713–723.

Further reading

Brewer, W., Castle, D. and Pantelis, C. (2012) *Olfaction and the Brain*. Cambridge: Cambridge University Press.
Colaianni, A. and Cook-Deegan, R. (2009) Columbia University's Axel patents: Technology transfer and implications for the Bayh-Dole Act. *Milbank Quarterly*, 87, 683–715.
Doty, R.L. (2015) *Handbook of Olfaction and Gustation*. London: Wiley-Blackwell.
Menini, A. (2010) *The Neurobiology of Olfaction*. Boca Raton: CRC Press/Taylor and Francis.
Mombaerts, P. (2004) Love at first smell – The 2004 Nobel Prize in Physiology. *New England Journal of Medicine*, 351, 2579–2580.

Chapter 34

Giacomo Rizzolatti (b. 1937)

Rizzolatti is famous for discovering mirror neurons in certain regions of the brain that fire when an animal executes a motor act while observing a similar act at the same time – a neural mechanism that appears to be involved in understanding other people's actions and emotional empathy.

Giacomo Rizzolatti was born in Kiev, Ukraine, in 1937, to an Italian father and Russian mother who were doctors. This was a period known as the Great Purge in the Soviet Union, characterised by widespread police surveillance, suspicion and arbitrary executions that resulted in the death of over a million people. There was also compulsory repatriation, and Rizzolatti was only 6 months old when his family were forced back to Italy where they settled in the north eastern city of Udine. The situation in Italy was also difficult for the country was on the brink of entering World War II, and Rizzolatti's father was soon recruited into the military as a medic and sent to the Yugoslavian front. Nor was Udine safe, for after its German occupation, the city was bombed by the Allies, and Rizzolatti would remember seeking refuge in air-raid shelters. But this did not deter the young boy from learning how to read or play chess, and by the time he went to school, Rizzolatti was in his own words "precocious". The end of the war in 1945, however, bought a marked change in fortune, with Rizzolatti's father becoming a general practitioner and earning good money. At the age of 13, Rizzolatti began his secondary school studies at the Ginnasio-Liceo Iscopo Stellini of Udine – one of the most prestigious schools in Italy, famous for its library and collection of valuable old books. Here, Rizzolatti was educated in classical studies including Latin and Greek, along with history, literature and philosophy. Passing his school exams at the age of 18, Rizzolatti moved some hundred miles south-west to the medical faculty at the University of Padua – one of the oldest medical schools in the world, with a long list of famous alumni including Nicolaus Copernicus, Andreas Vesalius and William Harvey. Yet, Rizzolatti found his medical studies rather boring – that is until his fifth year when he began to study neurology and attend patients with neurological disease. Even if he had not been over enthusiastic about all his subjects, Rizzolatti still graduated in 1961 with a MD and maximum marks of *cum laude*.

Over the next 7 years, Rizzolatti would hold a number of short-term research positions. While still serving his internship in neurology, Rizzolatti became proficient using the EEG and published his first paper on light-sensitive epilepsy. In 1963, Rizzolatti moved to the University of Ferrara where he gained practical experience in neurophysiology and undertook a study that recorded neural activity from the pyramidal tract (a path of motor neurons that originate in the cerebral cortex and terminate in the spinal cord). A year later and now fully qualified in neurology, Rizzolatti took up a post at the University of Pisa, one of the leading neurophysiological research institutes in Europe, where he investigated the visual properties of neurons in the lateral geniculate nucleus and superior colliculus of the cat. Rizzolatti also became interested in the role of the corpus callosum in the inter-hemispheric transfer of information, especially when Michael Gazzaniga arrived in his laboratory after working with Roger Sperry. These were, in effect, Rizzolatti's research apprenticeship years, and they led to a tenured post at the University of Parma in 1967 where he became professor within a few years.

In 1971, Rizzolatti began his first experiments on monkeys. One of the brain regions that interested him were the frontal eye fields – an area located in the frontal cortex, just in front of the premotor cortex. This area was known to be involved in the shifting of gaze and the generation of eye movements, especially when tracking a moving object. However, in his attempt to record from neurons in this area, Rizzolatti also came across an adjacent region located in the premotor region itself, which only responded to visual stimuli close to the body or within the monkey's reaching distance. This was an unexpected finding since this area of the brain was believed to be primarily involved in the planning and execution of motor action through its direct projections to the spinal cord. Rizzolatti also found that many of its neurons were bimodal, responding to both visual and tactile information, as occurs in hand-to-mouth movements. Clearly, the motor areas of the cortex still had secrets to give up, and in an attempt to elucidate their role, Rizzolatti began to map out its regions with new neuroanatomical techniques. According to Rizzolatti, this would show that the motor cortex does not consist of two main areas (e.g. primary and premotor areas) as classic accounts of the brain held but a mosaic of cortical areas with specific connections and functional properties. Rizzolatti would number these areas F1 to F5. Later, he would add two more regions (F6 and F7) to his classification of the premotor cortex.

In exploring the behavioural functions of his newly defined premotor regions, Rizzolatti obtained the most interesting results by recording from F5, which lay in the lateral-ventral part of premotor cortex – a site partly corresponding to Broca's area in humans. Here, he found the firing of single neurons correlated with specific goal-related motor acts rather than general bodily movements as had been shown by others. In fact, Rizzolatti identified four types of specialised neuron in area F5: (1) grasping-with-the-hand-and-mouth cells, (2) grasping-with-the-hand-only cells, (3) holding-in-hands cells and (4) tearing-up cells. Although all these types of neuron were associated with action, they also required visual stimuli to operate. For example, Rizzolatti showed that neurons discharging during a precision grip only became active while observing a small object such as a peanut, whereas neurons encoding a power grip only became active during the observation of a large object like an apple.

It was during these studies, in 1991, that Rizzolatti and his colleagues began to train monkeys to retrieve objects of different sizes and shapes from a testing box. After a few learning trials, Rizzolatti observed that a relatively large proportion of F5 neurons discharged when the monkey was observing the experimenter performing a specific motor act such as picking up a peanut. Moreover, these were the same neurons that fired when the monkey handled the peanut. Put simply, these neurons fired when the monkey executed an action and also fired when they saw the same action taking place in another individual. Soon after, Rizzolatti recognised that some neurons in F5 only fired when the monkey put a peanut in its mouth and would fire again when the experimenter repeated the action. In the winter of 1991, Rizzolatti submitted his findings to the reputable journal *Nature*, but the editor rejected the paper for its "lack of general interest". However, it was later published in *Experimental Brain Research* (Di Pellegrino et al. 1992) and has since become one of the most cited papers in neuroscience with over 2500 citations. Four years later, in an article in the journal *Brain*, Rizzolatti introduced the term *mirror neurons* – a phrase that neatly encapsulated the function of these imitatory cells and inspire countless researchers since.

At this point, mirror neurons were conceptualised as a highly specialised type of motor neuron, only found in the frontal cortex, that became activated when an individual performed a specific act that could then be repeated by the observer. These neurons were also specific in terms of their visual requirements since they did not fire in response to the simple presentation of food or other interesting objects. Nor did they discharge when the animal simply observed a hand action mimicked without the target object. Hence, both an action and object were needed to be together, or interacting, for the mirror neuron to show activity.

For many, the next big question was obvious: to see whether mirror neurons existed in the human brain. Clearly, Rizzolatti could not perform the same invasive single cellular recording studies as he had done in the monkey, and so a different approach was needed. The answer lay with functional brain imaging studies that Rizzolatti initiated in the mid-1990s. At first, he used PET imaging of regional blood flow, with subjects lying in the scanner while observing an experimenter handle various common objects, after which they had to reach out and grasp the same object. Since then, brain imaging of mirror neurons has become one of the most popular subjects in cognitive neuroscience, with the observation of film, rather than direct action, providing the choice method.[1] In 2010, a meta-analysis of 139 fMRI and PET mirror neuron experiments by Caspers et al. (2010) showed that three main regions of the human brain were involved in mirroring behaviour. These were: (1) the frontal cortex including the lateral premotor cortex and adjacent supplementary motor area, (2) the inferior parietal lobe and (3) the posterior middle temporal gyrus that borders on the occipital lobes. This work has been confirmed by fMRI studies on monkeys, which has also enabled Rizzolatti to trace the neural interconnectivity of these regions with standard anatomical techniques. According to Rizzolatti, this work has shown that a subset of areas in the temporal lobe passes information to the frontal cortex, along two major pathways, which involve different parts of the parietal lobes. One of these pathways is involved in recognising the agent of the motor act, while the other is concerned with the details of the hand grip or body movement.

In 2001, Rizzolatti greatly broadened the concept of mirror neurons when he proposed that they are not only involved in recognising the actions of others but also in *understanding* the goal of an observed behaviour. Thus, in his view, mirror neurons are not just capable of mimicry but are also in assessing *why* an action is taking place. This now strongly implicated mirror neurons as a vital component of intelligence along with many other types of cognate human behaviour. One obvious role for mirror neurons in this respect was imitation learning, which is particularly vital in child development, including play behaviour, the acquisition of language and social interaction. In fact, Rizzolatti would claim that, thanks to this mirror mechanism, overt actions undertaken by other individuals become messages that are intuitively understood by an observer without any cognitive mediation being necessary. This theory has generated much research and several findings support the idea. For example, it has been demonstrated that human infants as young as 2 weeks old will imitate an adult who sticks their tongue out at them (Meltzolf and Moore 1977), a finding that appears to back up Rizzolatti's claim.

Another interesting development has been the demonstration that mirror neurons are also important for recognising and understanding emotion – otherwise known as empathy. For example, in one fMRI study, subjects were introduced to an odor of rotten eggs that caused activation of neurons in the anterior insular cortex as well as cingulate cortex. When the subjects were then shown images of disgusted faces, this too produced activation of the insular cortex (Wicker et al. 2003). Similarly, observing or imaging other people's pain has been shown to activate neural circuits known to be involved in the experience of pain itself. According to Rizzolatti, this explains why you "feel" an actor's emotion or start smiling in response to a happy person. Empathy is an emotion that many psychologists believe is not just crucial for successful human interaction but also defines us as human beings. If Rizzolatti is correct, then this vital component of human social behaviour is something we don't have to think about – we automatically react by experiencing another person's emotion. Not surprisingly, this work has also generated considerable interest in determining whether dysfunction of mirror neuron circuits in the brain may underlie disorders such as Autism where empathy and social competence are compromised.

Rizzolatti's identification of mirror neurons has become one of the most exciting areas of research in recent years and has put the problem of how we understand others at the forefront of social neuroscience. For many, mirror neurons represent a basic principle of brain function and are found in many more brain regions than was first envisaged. A particularly strong advocate of mirror neurons has come from the eminent Indian neurologist V.S. Ramachandran who wrote in 2000: "I predict that mirror neurons will do for psychology what DNA did for biology: they will provide a unifying framework and help explain a host of mental abilities that have hitherto remained mysterious and inaccessible to experimenters ... With knowledge of these neurons, you have the basis for understanding a host of very enigmatic aspects of the human mind ... (including) mind reading empathy, imitation learning and even the evolution of language". Ramachandran also believes that the development of mirror neurons is the driving force behind the great leap forward in human evolution. These are bold statements, although they are not shared by all researchers. In fact, some sceptics such as Gregory Hickok at the University of California, Irvine (who has written *The Myth of Mirror Neurons*), believe that mirror neurons are "monkey business" and do not explain in any meaningful way the intricacies of human behaviour including imitation and empathy. Others point out that although we may have mirror neurons in our brain, along with other animals, this does not necessarily prove that they are "reading" another's mind – be it a monkey or a human. Although the controversy continues, one thing seems certain: research into mirror neurons shows no signs of abating and will continue for a long time yet.

Note

1 Other methods that have been used to examine mirror neurons are EEG, magnetoencephalography and transcranial magnetic stimulation

Rizzolatti's major writings

Di Pellegrino, G., Fadiga, L., Fogassi, L., Gallese, V. and Rizzolatti, G. (1992) Understanding motor events: A neurophysiological study. *Experimental Brain Research*, 91, 176–180.

Fogassi, L., Ferrari, P.F., Gesierch, B., Rozzi, S., Chersi, F. and Rizzolatti, G. (2005) Parietal lobe: From action organization to intention understanding. *Science*, 308, 662–667.

Rizzolatti G, Fogassi L. and Gallese V. (2001) Neurophysiological mechanisms underlying the understanding and imitation of action. *Nature Reviews. Neuroscience* 2, 661–670.

Rizzolatti G. and Craighero, L. (2004) The mirror-neuron system. *Annual Review of Neuroscience*, 27, 169–192.

Rizzolatti, G. and Sinigaglia, C. (2007) *Mirrors in the Brain: How our Minds Share Actions, Emotions and Experience*. Oxford: Oxford University Press.

Rizzolatti, G., Cattaneo, L., Fabbri-Destro, M. and Rozzi, S. (2014) Cortical mechanisms underlying the organization of goal-directed actions and understanding. *Physiological Review*, 94, 655–706.

Wicker, B., Keysers, C., Plailly, J., Royet, J.P., Gallese, V. and Rizzolatti, G. (2003) Both of us disgusted in my insula: The common neural basis of seeing and feeling disgust. *Neuron*, 40, 655–664.

Further reading

Caspers, S., Zilles, K., Laird, A.R. and Eickhoff, S.B. (2010) ALE meta-analysis of action observation and imitation in the human brain. *Neuroimage*, 50, 1148–1167.

Ferrari, P.F. and Rizzolatti, G. (2015) *New Frontiers in Mirror Neuron Research*. Oxford: Oxford University Press.

Hickok, G. (2014) *The Myth of Mirror Neurons: The Real Neuroscience of Communication and Cognition*. New York: Norton and Co.

Meltzoff, A.N. and Moore, M.K. (1977) Imitation of facial and manual gestures by human neonates. *Science*, 198, 75–78.

Ramachandran, V.S. (2000) Mirror neurons and imitation learning as the driving force behind "the great leap forward" in human evolution. www.edge.org/3rd_culture/ramachandran/ramachandran_p1.html.

Singer, T. (2006) The neuronal basis and ontogeny of empathy and mind reading: Review of literature and implications for future research. *Neuroscience and Biobehavioral Reviews*, 6, 855–863.

Chapter 35

John O'Keefe (b. 1939)

O'Keefe is a Nobel winning neuroscientist, who is famous for his discovery of place cells in the hippocampus – neurons that encode an animal's location in its geographic space and which are involved in navigation and creating a cognitive map of the environment.

John O'Keefe was born in Harlem, New York, to Irish immigrant parents – neither of whom had completed elementary education, although his father would later attend evening classes and attain a high school degree. Growing up in Bronx, O'Keefe's father worked as a nightshift mechanic on the New York bus system and his mother a welder in the Newark shipyards during the war years. Despite this unpromising start in life, the young boy showed academic promise and won a scholarship to a prestigious private Jesuit school, Regis High, in Manhattan. Unfortunately, not coping well with its diet of Latin and Greek, O'Keefe left school at the age of 16 with poor grades, taking clerical jobs in a Wall Street stock brokerage house and then an insurance company. But office work did not fulfil O'Keefe's ambitions and he started an evening course in aeronautical engineering at New York University while working during the day. This would lead to more suitable employment on Long Island, although the gruelling commute of over 100 miles in rush hour traffic combined with the long hours of his evening study took its toll. So in 1960, at the age of 21, O'Keefe enrolled full-time at City College of New York where he studied a wide range of courses – opting to major in psychology with a minor in philosophy. This also resulted in O'Keefe becoming introduced to physiological research when he examined the effects of brain lesions on the exploratory behaviour of pigeons. Graduating in 1963, O'Keefe moved to McGill University in

Montreal to undertake his PhD under the guidance of pain expert Ronald Melzack. At the time, Melzack had just received a large grant to build an electrophysiological laboratory, and O'Keefe used the new equipment to develop techniques for recording from single cells in the rat's amygdala (this entails implanting fine wires extracellularly but which lie close enough to a cell to pick up its electrical activity). Soon after obtaining his doctorate in 1967, O'Keefe took up a research position at University College, London, and quickly fell in love with the landscape and culture of Britain. Determined to stay but without a tenured position, O'Keefe was at the mercy of UK research bodies who supplied the grants, enabling him to continue working.

In 1970, O'Keefe began his first studies with the hippocampus, and as with so many new breakthroughs, this development was more a case of good fortune than planned design. At the time, O'Keefe was examining the relationship between neural activity in the motor regions of the cerebral cortex and spinal cord, and in one set of experiments, he attempted to implant a microelectrode in the somatosensory region of the thalamus but placed it too laterally so it strayed into the hippocampus. He was immediately struck by the close correlation of the theta rhythmicity of the hippocampal neurons (theta activity occurs when the cell fires in oscillations that have a frequency of 4–7 Hz) and the rat's movement, and he decided at this point to forget the somatosensory system and begin examining the hippocampus. O'Keefe was also intrigued by the apparent difference between the motor functions of the hippocampus, which he had just witnessed, and its proposed role in memory as had been shown in patient HM and many other studies.

Around this time, a graduate student from the Israel Institute of Technology in Haifa called Jonathan Dostrovsky joined O'Keefe's laboratory to undertake a master's degree. Together they began recording from single hippocampal cells while the rat was engaged in a variety of basic behaviours including eating, drinking, grooming, exploring and learning tasks such as lever pressing. They soon realised the hippocampus contained two types of cells that were distinguishable by their electrophysiological properties. The first type showed random bursts of activity when the animal was sitting quietly or in a quiescent state. The second produced theta activity when the rat moved, and this was most marked when the animal changed location. At first, O'Keefe did not realise the significance of the theta observation, until some months later "on one electrifying day", he insightfully recognised that these cells were responding to the exact site where the animal was located in its environment. This was verified when O'Keefe and Dostrovsky placed the rat on a raised platform surrounded on three sides by a white plastic curtain and found that while a single hippocampal neuron might fire twenty times a second when the rat was in a particular location, it only did so intermittently in another place. Moreover, when the curtain surrounding the platform was removed, the cell's activity ended abruptly, indicating that these cells were firing in response to the external (extra-maze) cues of its environment. O'Keefe published the results in a short three-page paper (O'Keefe and Dostrovsky 1971) where he introduced the term *place cells* and speculated that the hippocampus acted to provide the rest of the brain with a spatial reference map.

Around the same time, an old American friend from McGill called Lynn Nadel, who had fled his postdoctoral fellowship in Prague after the Russian invasion of Czechoslovakia in 1968, joined O'Keefe's laboratory. Although Nadel had intended to work on the visual system, he quickly became interested in place cells and set about writing a short review article on them. The result after many tribulations lasting 6 years would be a book of 570 pages that included over 1500 references entitled *The Hippocampus as a Cognitive Map* (1978). It provided a strong argument for the theory that the hippocampus and its associated brain regions produced spatial computations and created geographical maps that enabled navigation. In addition, O'Keefe and Nadel suggested that similar hippocampal neural mechanisms might underlie other types of cognition including episodic memory and even the deep structure of language. Although these claims were speculative, a number of studies have indeed since implicated the hippocampus in several different types of cognition.

The key feature of the cognitive mapping theory concerned place cells, and by the late 1970s, O'Keefe had begun to realise that these neurons must be receiving and utilising two types of specialised information. The first type of information was the spatial cues, or landmarks, which provide the 3-D frame of reference for our external world. But, just as important, in order to navigate through this spatial layout, the hippocampus also had to compute other types of information, such as the direction and speed of locomotion, etc. – a calculation that uses the physical map but is independent of it. In fact, this process has been likened to the method of "dead reckoning" used by navigators, which requires fixing the current position and then heading towards the goal by taking a bearing and estimating the speed and distance to be travelled.

Convincing evidence that hippocampal place cells utilise landmark-based information that is dependent on the array of spatial cues came from a study O'Keefe performed with research student Dulcie Conway. They trained rats on a T-maze that was surrounded by a black curtain that had an object attached to each of the four sides (a fan, buzzer, light and a square card). The curtain could also be moved around the maze, with the cues changing location but always remaining in the same configuration to each other. The results showed that the place cells only fired when the animal was in a specific location relative to the cues. If the curtain was moved around, the place cells no longer fired, unless the rat was moved to the same relative position to the cues. The importance of the extra-maze cues was also confirmed when they were removed. If two cues were taken away, the place cells tended to keep firing but not if three cues were removed. Clearly, therefore, place cell activity is dependent on a spatial configuration of different landmarks and not any single object or cue.

But perhaps the strongest evidence for hippocampal involvement in cognitive mapping came from an experiment by Richard Morris, who had close ties with O'Keefe's laboratory. In the early 1980s, he developed a task where rats were required to swim in a water maze. The maze was in fact a large circular tank of water made opaque with the addition of milk, which also contained a small escape platform whose surface was hidden just below the water level. Rats were put into the water and allowed to swim around until they "bumped" into the platform – onto which they inevitably climbed. The rats could not directly see the platform and were forced to learn its location by determining its position in relation to the external cues around the maze. Morris found it took just a few trials for normal animals to consistently and quickly find the platform. In contrast, the hippocampal lesioned rats were unable to learn the location of the platform after 40 days of training. As O'Keefe has acknowledged, the Morris water maze is both a superb test of spatial navigation and hippocampal function.

The cognitive mapping theory received further support when various cells involved in directional "dead reckoning" types of computation were identified. The first of these were discovered in the early 1980s when O'Keefe, along with Bruce McNaughton and Carol Barnes, who were visiting postdoctoral students to his laboratory, found place cells in the hippocampus whose firing rate was dependent on the animal's speed of movement. That is, the faster the speed, the more the neuron fired. Moreover, many of these cells also had a movement bias – becoming active when the rat moved in one particular direction. Soon after this, O'Keefe would also go on to find a small number of "pure speed cells" in the hippocampus that fired irrespective of the animal's location or its direction of movement. In New York, another team of investigators led by Jim Ranck would show cells in the subiculum (an area adjacent to the hippocampus) that specifically signalled the animal's head direction. Intriguingly, these head direction cells could also become "confused" when the rat was spun around – so that a cell that had an original north-east bias might now be altered to point in an eastward direction. And a third type of cell, called "boundary cells" by O'Keefe, was found during the 1990s in several regions of the hippocampal formation, which only responded to an environmental boundary at a particular distance and direction from the animal.

But the question of how the hippocampal system encoded distance remained unanswered. That is, until the husband and wife team of May-Britt and Edvard Moser, working at the Norwegian University of Science and Technology in Trondheim, Norway, found the answer. The pair had married as undergraduate psychology students, and despite having two children, both completed their PhDs at the University of Oslo in 1995. As part of their postdoctoral research, they worked in Edinburgh with Richard Morris, and in 1996, they spent time working with O'Keefe in London, where they learned how to perform single cell recordings from the hippocampus – an experience that May-Britt Moser described as one of the most learning-rich periods in her life (Edvard had 3 months working with O'Keefe and Brit-May just one, largely because of the difficulty of finding care for the children). After going back to Norway, they began examining the differences in function between the dorsal and ventral hippocampus. And in 2005, as they looked at other areas of the hippocampal system, they discovered a type of neuron in the nearby medial entorhinal cortex, which they referred to as *grid cells*. These were unique because they fired at multiple locations in the environment, in a highly regular equidistant pattern as the animal moved across the floor. It was as if a beautiful symmetrical pattern of equilateral triangles had been drawn onto the floor, and the grid cell was repeatedly firing in each triangle as the rat moved through each one. Moreover, this firing pattern was not influenced by the spatial configuration of landmarks and even occurred when the animal moved in the dark. Nor were grid cells influenced by changes in the animal's speed or direction. In short, the rigid map encoded by the grid cells appeared to provide a coordinate system by which distances could be measured, which were necessary for spatial navigation. It should also be noted that this neural system is the only one known to be generated entirely in the brain, marking it an important milestone in the history of computational neuroscience.

In 2014, John O'Keefe, along with May-Britt and Edvard Moser, were awarded the Nobel Prize "for their discoveries of cells that constitute a positioning system in the brain". In announcing the award, the Nobel Assembly noted that O'Keefe and his colleagues had not only solved a problem that had occupied philosophers and scientists for centuries but had also opened up a new paradigm for exploring many other cognitive processes including memory, thinking and planning. Among his many other awards, O'Keefe was elected fellow of the Royal Society in 1992 and of the Academy of Medical Sciences in 1998. He was also elected to the National Academy of Sciences in 2016. After spending so much of his career as a "bench experimenter", O'Keefe was thrilled when offered the inaugural directorship of the Sainsbury Wellcome Centre for Neural Circuits and Behaviour at University College, London, in 2013. Although he would step down from the directorship in 2016, O'Keefe continues to be actively engaged there as a group leader, studying the neural correlates of cognition and emotion using the latest techniques in optical and electrophysiological recording. He is also optimistic that an understanding of the hippocampus will allow us to find out more about Alzheimer's disease, and to this end, is intent on developing mouse models of place cell function to show how hippocampus physiology can become dysfunctional during disease progression. Sadly, May-Britt and Edvard Moser announced they would divorce in 2016.

O'Keefe's major writings

Morris, R.G., Garrud, P., Rawlins, J.N. and O'Keefe, J. (1982) Place navigation impaired in rats with hippocampal lesions. *Nature*, 297, 201–206.

O'Keefe, J. and Dostrovsky, J. (1971) The hippocampus as spatial map: Preliminary evidence from unit activity in the freely moving rat. *Brain Research*, 34, 171–175.
O'Keefe, J. (1976) Place units in the hippocampus of the freely moving rat. *Experimental Neurology*, 51, 78–109.
O'Keefe, J. and Conway, D.H. (1978) Hippocampal place units in the freely moving rat: Why they fire when the fire. *Experimental Brain Research*, 31, 573–590.
O'Keefe, J. and Nadel, L. (1978) *The Hippocampus as a Cognitive Map*. Oxford: Clarendon Press.
O'Keefe, J., Burgess, N., Donnett, J.G., Jefery, K.J. and Maguire, E.A. (1998) Place cells, navigational accuracy, and the human hippocampus. *Philosophical Transactions of the Royal Society of London, B*, 353, 1333–1340.
O'Keefe, J. (2014) Spatial cells in the hippocampal formation. www.nobelprize.org/nobel_prizes/medicine/laureates/2014/okeefe-lecture.html.
O'Keefe, J. (2014) John O'Keefe – Biographical. www.nobelprize.org/nobel_prizes/medicine/laureates/2014/okeefe-bio.html.

Further reading

Burgess, N. (2014) The 2014 Nobel Prize in Physiology or Medicine: A spatial model for cognitive neuroscience. *Neuron*, 84, 1120–1125.
Etienne, A.S. and Jeffery, K.J. (2004) Path integration in mammals. *Hippocampus*, 14, 180–192.
Hafting, T., Fyhn, M., Molden, S., Moser, M.B. and Moser, E.I. (2005) Microstructure of a spatial map in the entorhinal cortex. *Nature*, 436, 801–806.
Kumaran, D. and Maguire, E.A. (2005) The human hippocampus: Cognitive maps or relational memory? *The Journal of Neuroscience*, 25, 7254–7259.
McNaughton, B.L., Battaglia, F.P., Jensen, O., Moser, E.I. and Moser, M.B. (2006) Path integration and the neural basis of the "cognitive map". *Nature Reviews Neuroscience*, 7, 663–678.

Chapter 36

Michael Gazzaniga (b. 1939)

Gazzaniga is best known for his work on split-brain patients, which has spanned more than 50 years, his interpreter theory of left hemispheric function and for being one of the founders of cognitive neuroscience – a term he coined in the back of a New York taxi in the late 1970s.

Michael Saunders Gazzaniga was born in Los Angeles, California, and grew up in Glendale, the fourth of five children. His father, Dante Achilles Gazzaniga, was of Italian descent, and a well-known surgeon, who had become a founder of the Ross-Loos Medical Group – the first health maintenance organisation in the United States to provide a prepaid health services plan. Although a domineering figure totally committed to medicine (the patient always came first), Gazzaniga's father often had multiple activities on the go – one of which was to run biochemistry experiments in the garage with Michael. His mother was one-time secretary to the evangelist Aimee Semple McPherson and "the original Californian girl" who enjoyed putting on dinner parties and having friends in the house. It was in Gazzaniga's own words "a no-nonsense sort of child raising" with lots of fun things going on, despite the emphasis on hard work. After graduating from Glendale High School, Gazzaniga won a place at Dartmouth College, a private Ivy League university in New Hampshire, where his older brother Alan was already a star football player about to enter medical school. Gazzaniga soon fell in love with the place, joining the Alpha Delta Phi fraternity (the inspiration for the movie "Animal House") and discovering Baker Library with its famous Tower Room. Mesmerised by his new life, Gazzaniga also began taking an interest in psychology and brain science, working in the laboratory of

William M. Smith who was an expert of visual feedback and motor behaviour. During his third year, Gazzaniga came across a *Scientific American* article on Roger Sperry's nerve growth experiments performed at the Californian Institute of Technology (Caltech). Soon after, with his girlfriend planning to go back home to Los Angeles that summer, Gazzaniga wrote to Sperry asking if he would take him on as an intern. Gazzaniga got the job, although he was introduced to a new field of research – namely, the behavioural function of the corpus callosum. In fact, Gazzaniga spent much of that summer trying to find a way of reversibly anaesthetising one half of a rabbit brain. Gazzaniga was smitten with the work and returning to Dartmouth began seriously considering what sort of experiments he would run if he ever got access to humans with callosum lesions.[1] It was an obsession that did not leave him, and after obtaining his BA in 1961, Gazzaniga gave up his intention to pursue medicine (despite his father's disappointment) and head back to Caltech.

The question of corpus callosum function in humans was one of the great puzzles of brain research in the early 1960s. During the 1940s, the surgeon William van Wagenen, working at the University of Rochester in upstate New York, had severed the corpus callosum in a group of twenty-eight patients suffering from severe and intractable epilepsy. The intention was to confine the electrical storm of the seizure to just one side of the brain – a surgical procedure that led to an improvement in only a handful of the patients. The behavioural effects of the surgery were followed up by psychiatrist Andrew J. Akelaitis, who was an expert in cognitive and language function. Curiously, he found that the corpus callosotomy did not noticeably affect thought, emotion or behaviour in the patients. Considering the prominence of the corpus callosum that connected the two large cerebral hemispheres, the results were surprising. They even led Karl Lashley to quip in 1951 that the only known function of the corpus callosum was to stop the hemispheres from sagging! But in the 1950s, Sperry began to show otherwise when he began training commissurotomised cats on various tasks such as maze learning, with eye patches used to train individual hemispheres. His work showed that the corpus callosum was clearly involved in the inter-hemispheric communication of information.

Around the same time, a neurosurgeon at Caltech, Joseph Bogen, was reviewing the earlier studies by Van Wagenen and reopening the question of whether callosotomy could benefit epileptic patients.

In 1962, he enlisted the help of Philip Vogel to perform their first commissure sections and asked Sperry if he would like to test the patients afterwards. As Gazzaniga recounts, as soon as he arrived for his graduate research, Sperry gave him the assignment. The first day of testing came on a bright sunny Pasadena day in 1962 and Gazzaniga's first patient was a former paratrooper in World War II, with the initials WJ, who had become epileptic after being hit over the head with a rifle butt. Gazzaniga had already come up with a clever procedure by which to project visual information to a single hemisphere. In humans, the visual system is crossed, with each hemisphere receiving input from both eyes. But if a stimulus is projected laterally enough into the visual field of one of the eyes, then it will pass to the opposite-sided hemisphere. Hence, Gazzaniga got his patient to fixate on a small spot in the centre of a screen and then flashed simple images for one-tenth of a second to each side. This was sufficient time for the input to reach one of the hemispheres but not enough for the eyes to turn and fully focus on the image, making it bi-hemispheric. Later, Gazzaniga used another technique to present the hemispheres with tactile information. He did this by blindfolding patients and getting them to identify an object placed in one of their hands. In humans, the pathways conveying touch from the body, unlike the visual system, are fully crossed. Hence, when patients were asked to hold an item in their right hand, the tactile information passed to the left hemisphere and vice versa. In much the same way, the right hemisphere controls the left hand and the left controls the right.

It soon became clear that the two hemispheres had different functions. When Gazzaniga used a board of lights as a means of presenting visual stimuli, those presented to the left visual field (right hemisphere) elicited no response, whereas those flashed into the right visual field (left hemisphere) were correctly reported. Despite this, the left hand (controlled by the right hemisphere) could accurately point to the light source. Thus, only the left hemisphere could verbalise what it had just observed. Gazzaniga confirmed this by presenting written words to the hemispheres. If a word was presented to the left hemisphere, the subject could read it and understand its meaning. But the same word to the right hemisphere typically elicited no response. Despite this, the right hemisphere was not without its linguistic abilities. For example, when an object name (e.g. *spoon*) was flashed to the right hemisphere, the left hand could select it from a range of objects hidden behind a screen.

However, in this case, the subject typically could not describe what they had picked up. Remarkably, in one experiment where different object names were flashed simultaneously to both hemispheres (e.g. *key-ring*), it was found that patients typically picked up the object presented to their right hemisphere (*ring*) but name the object given to the left hemisphere (*key*). The subjects were also astonished to find out that they had misnamed the object when it was bought into full view. Gazzaniga would also show that the right hemisphere was far better than the left in arranging blocks to match a pictured design or draw a cube in three dimensions.

Gazzaniga's remarkable experiments were first published in 1962, and he would go on to examine ten other patients with a variety of tests. Although Sperry sometimes sat in on the testing, Gazzaniga was the "chief operating officer" who designed and ran the studies. Yet, after obtaining his PhD in 1965, there was no place for Gazzaniga at Caltech, and he would leave to spend a year in Pisa, where he recorded from neurons in the corpus callosum and made friends with Giacomo Rizzolatti. Almost 20 years on from Gazzaniga's first work with WJ, Sperry was awarded a Nobel Prize in 1981 for his work on split-brain patients. Gazzaniga would describe his time at Caltech as "unforgettable and enduring" and Sperry's Nobel Prize as "fully deserved". Yet, Gazzaniga would not take to neurophysiology and returned home in 1968 to take up a post at UC Santa Barbara – an institution only 100 miles from Caltech. But access to his split-brain patients had become less viable[2] and within a year Gazzaniga moved east. For the next 20 years, he would be based in universities in New York.

In New York, Gazzaniga managed to access more split-brain patients, although much of his work now focused on the unconscious processing of information in other types of patients. For example, he examined *blindsight* (first described by Larry Weiskrantz in the early 1970s), where "blind" patients with visual cortex damage, can accurately point to objects. It was believed that this ability depended on subcortical systems that relayed the unconscious visual input to the conscious cerebral hemispheres. But Gazzaniga's work using an image stabiliser, whereby a stimulus is projected into the blind part of the eye, cast doubt on this. In fact, Gazzaniga showed that blind regions of the eye often had small islands of vision, although the subject still reported being blind – and these also could explain blindsight. Another deficit Gazzaniga explored was *contralateral neglect*. This occurs after right-sided parietal lobe damage, which causes the person to ignore the left side of their world. But they are not blind, as shown when asked to copy a clock face, which can result in all twelve numbers being drawn on the right. Gazzaniga confirmed that the left side of the world was being unconsciously processed in these patients by asking them to judge whether two visual stimuli, one appearing in each visual field, were the same or different. Thus, a patient might see an apple in their right visual field and an orange in the left one. Although the subject denied seeing the stimulus presented in their left visual field, they could still perform the task.

In the early days of split-brain research, it was widely believed that each hemisphere had its own specialised mind. However, in recent years, Gazzaniga has come to a different view that attributes the left hemisphere with "an interpreter" that attempts to understand the totality of what is happening to the individual. For example, in one study, a split-brain subject (PS) was presented with two unrelated pictures: the right hemisphere was shown a winter landscape with snow and the left a chicken's claw. After seeing these two pictures, PS was given an array of new pictures and each hemisphere required to pick one that went with the first presentation. Gazzaniga found that the two hemisphere selected appropriate pictures, with PS's left hand (right hemisphere) pointing to the shovel and his right hand (left hemisphere) to a chicken. When PS was asked why he had chosen the shovel, his speaking left hemisphere replied that it could be used to clean out the chicken house. Yet, the left hemisphere had no way of knowing why the right hemisphere had chosen the shovel. According to Gazzaniga, this shows that the left hemisphere has an interpreter that "tries to bring order and unity to our conscious lives" even when it has incomplete information.

In another study, Gazzaniga used an elaborate optical system on a female split-brain patient (VP), which projected a movie exclusively into the left visual field. It was also programmed so if VP moved her eyes towards the movie, the projector turned off. The right hemisphere then saw a film where an aggressor pushed another man off a balcony and then tried to set him alight. When VP was asked about the movie, she was unable to report any details. Despite this, she reported feeling nervous and said to one of the assistants "I know I like Dr Gazzaniga, but right now I am scared of him for some reason". As Gazzaniga notes, this happens to all of us. Our mental systems set up an emotion that alters the state of the brain, and in response, the interpreter recognises the mood, thereby attributing a cause to the feeling based on all the available evidence.

The theory of the interpreter is a beguiling concept. It argues that both hemispheres contain specialised modules for performing a wide range of mental functions. These are not organised hierarchically (there is no overall controller as such) but operate in a parallel and distributed system. Gazzaniga also believes that decision-making, judgement, perception and almost every other cognitive and emotional actions in the brain are unconscious – and what we recognise as consciousness arises after in the left hemisphere, through a process which Gazzaniga calls "post-hoc rationalisation". Thus, the interpreter is the part of the brain that pulls all the information together to provide meaning to our reactions. It is, in effect, creating a theory to justify what it has just experienced.

Michael Gazzaniga is currently professor of psychology at UC Santa Barbara, where he heads the SAGE Center for the Study of the Mind, and despite approaching his eighties, he shows no signs of slowing down. But perhaps his most lasting legacy to neuroscience occurred in a back of a New York taxi cab in the late 1970s when he and psychologist George Miller were en route to a dinner meeting at the Algonquin Hotel, which had been organised for scientists joining forces to study how the brain enables the mind – and they came up with the term *cognitive neuroscience*. Neither man remembers, or is willing to say, which of them first spoke the fateful name, but it took hold. Since then, Gazzaniga has helped found the Cognitive Neuroscience Society in 1994, which now has 2000 members worldwide, and the *Journal of Cognitive Neuroscience*. Gazzaniga's long experience working with split-brain patients has given him a unique perspective, which in recent years had led him to write on some of the wider questions concerning human behaviour, such as moral responsibility, free will and consciousness. He has also published many books that are accessible to a lay audience, while having a tribute volume dedicated to him in 2010.

Notes

1 In fact, Gazzaniga nearly did get access to a famous group of split-brain patients in Rochester, New York, who had received callosum lesions in the 1940s. After being promised access, Gazzaniga had

managed to rent a car that was loaded with equipment – only for the young undergraduate to be thwarted by a change of decision at the last minute.

2 In *Tales from Both Sides of the Brain* (page 116), Gazzaniga writes: "Sperry had made it increasingly clear through various surrogates that he didn't want me testing the Caltech patients anymore".

Gazzaniga's major writings

Gazzaniga, M.S. (1967) The split brain in man. *Scientific American*, 217, 24–29.

Gazzaniga, M.S. (2000) Cerebral specialization and interhemispheric communication. *Brain*, 123, 1293–1326.

Gazzaniga, M.S. (2008) *Human: The Science behind What Makes Your Brain Unique*. New York: Harper Perennial.

Gazzaniga, M.S. (2011) *Who's in Charge? Free Will and the Science of the Brain*. London: Robinson.

Gazzaniga, M.S. (2013) Shifting gears: Seeking new approaches for mind/brain mechanisms. *Annual Review of Psychology*, 64, 1–20.

Gazzaniga, M.S., Ivry, R.B. and Mangun, G.R. (2013) *Cognitive Neuroscience: The Biology of the Mind*. New York: Norton and Co.

Gazzaniga, M.S. (2015) *Tales from Both Side of the Brain: A Life in Neuroscience*. New York: Harper and Collins.

Gazzaniga, M.S. (2018) *The Consciousness Instinct: Unravelling the Mystery of how the Brain Makes the Mind*. London: Farrar Straus and Giroux.

Further reading

Gainotti, G. (2012) Unconscious processing of emotions and the right hemisphere. *Neuropsychologia*, 50, 205–218.

Reuter-Lorenz, P.A., Baynes, K., Mangun, G.R. and Phelps, E.A. (2010) *The Cognitive Neuroscience of Mind: A Tribute to Michael S. Gazzaniga*. Cambridge, MA: MIT Press.

Uddin, L.Q., Rayman, J. and Zaidel, E. (2005) Split-brain reveals separate but equal self recognition in the two cerebral hemispheres. *Consciousness and Cognition*, 14, 633–670.

Wolman, D. (2012) A tale of two halves. *Nature*, 483, 260–263.

Zaidel, E. and Iacoboni, M. (2003) *The Parallel Brain: The Cognitive Neuroscience of the Corpus Callosum*. Cambridge, MA: MIT Press.

Chapter 37

Joseph LeDoux (b. 1949)

LeDoux is best known for his research on fear conditioning, which has allowed him to map out the neural pathways in the brain involved in emotion and the reaction to dangerous events – work that shows a pivotal role for the amygdala.

Joseph Edward LeDoux Jr was born in Eunice, a small town in Louisiana's Cajun country. His father had been a travelling rodeo performer in his teens, before settling down to farming and establishing a butcher's business. Nonetheless, much to his father's disappointment, his son was not interested in riding horses or becoming a cowboy. Nor was LeDoux attracted by the prospect of being a butcher – although as an 8- or 9-year-old, he helped out in the family's business, removing the tough fibrous outer meninges from cow's brains and digging his fingers around in the soft flash so he could remove the bullet that had dispatched the animal in the slaughterhouse. As LeDoux has noted there is something unique about being in the presence of a brain compared to other bodily organs, which makes one contemplate what the mind must have experienced in life and also in the moments before death. Attending the devoutly Catholic St Edmund's Elementary School where he was taught by nuns, LeDoux considered the possibility of priesthood – that is, until puberty kicked in, when his interest shifted to the teenage passions of girls and guitars. Graduating from high school at the age of 18, LeDoux sought out university education. Although the Louisiana State University (LSU) had a campus in Eunice, LeDoux was desperate to leave home and managed to persuade his parents that the main campus in Baton Rouge would deliver a better education. They finally agreed, provided LeDoux took a major in business administration and promised to return to Eunice as a banker. Uninterested in business, LeDoux took courses in consumer psychology and marketing, with the result that he minored in psychology. In 1972, LeDoux continued his education at LSU by taking a master's degree in marketing. Again, he took psychology courses – one of which was taught by Robert Thompson, a well-known brain scientist, whose research interest concerned the neural basis of classical conditioning and memory. LeDoux had not even considered it remotely possible the brain could be studied, and so taken by this new realisation, he volunteered to work in Thompson's laboratory. After his master's degree, LeDoux looked around for a PhD placement and would apply for some twelve positions before managing to get accepted, in 1974, at the State University of New York at Stony Brook. Here, working with the split-brain patients of Michael Gazzaniga, LeDoux examined the ways in which the conscious and verbally dominant left hemisphere was able to deal with the unconscious "disconnected" behaviours of the right hemisphere (as occurs, for example, when the right hemisphere is given a command to stand, and the left hemisphere not knowing the reason confabulates an explanation, such as having to stretch the legs). This would be the start of LeDoux's interest in how non-conscious mental processes can act to influence conscious behaviour.

After completing his PhD in 1977, LeDoux joined the medical school at Cornell University as postdoctoral fellow and remained there until 1989, reaching the rank of associate professor. By now, LeDoux was becoming interested in how emotions generate neural signals in the brain, which are consciously interpreted as feelings – a curiosity that had been triggered by Gazzaniga over drinks one day, when he had remarked "Gee, there's not much research on emotion out there". However, because LeDoux could figure out no good way of examining emotion in humans, he decided to turn to rats, using some of the conditioning approaches learned under Robert Thompson. More specifically, LeDoux decided to make use of a procedure called "Pavlovian fear conditioning", whereby a neutral stimulus (e.g. a tone) is paired with an aversive stimulus such as a mildly aversive electric shock. By pairing these two stimuli together, it only takes a few trials for the animal to show a conditioned emotional response to the tone alone – as shown by the elicitation of several "fearful" sympathetic changes including increased blood pressure and sudden "freezing" of behaviour. LeDoux was particularly keen in understanding the neural networks of the brain that make it possible for an acoustic stimulus to produce this type of defensive fear reaction.

To begin, LeDoux traced the sensory neural pathway by which the tone elicited the behavioural fear response. This was relatively straightforward as the pathways from the ears to the brain were well known, with the auditory nerve projecting to the brain stem's cochlear nucleus and inferior colliculus of the midbrain. From here, sound information reaches the primary auditory cortex (located in the temporal lobe) via the medial geniculate nucleus of the thalamus.[1] LeDoux then set about identifying the important parts of this system for fear conditioning. One area to be lesioned was the auditory cortex – an operation that did not impair the fear response to the tone. He then destroyed the next lower station, the medial geniculate nucleus – a lesion that now impaired the rat's ability to learn the conditioning procedure, with the tone and shock pairings having little effect on blood pressure or the elicitation of freezing.[2] Clearly, this part of the thalamus was crucial for learning about fearful events, although it also raised a difficult question: if the tone stimulus was not reaching the auditory cortex, then where in the brain was it going?

To address this issue, LeDoux undertook a neuroanatomical tract tracing study by injecting into the medial geniculate nucleus a chemical called" wheat-germ agglutinin" – a substance taken up into the cell bodies and then transported down their axons to the synaptic terminals. Hence, by following the flow of wheat-germ agglutinin by using a special chemical reaction, LeDoux was able to see what areas of the brain, other than the auditory cortex, received projections from the medial geniculate nucleus. One structure stood out: a part of the limbic system in the anterior portions of the temporal lobe named by early anatomists for its almond nut shape – the *amygdala*. This is a small but complex nucleus consisting of at least twelve regions each with subregions. And the medial geniculate nucleus projected primarily to one of its areas – the lateral amygdala. It was then a small step to show that lesions of the lateral amygdala, like those of the medial geniculate nucleus, abolished the ability of the animal to learn the conditioning procedure.

The next question was to ask how the lateral amygdala produces its effects on fear-related behaviour. It was clear from other neuroanatomical studies that the lateral amygdala not only received sensory input from the thalamus but also from other areas of the brain including widespread regions of the neocortex and hippocampus. It was also apparent that the lateral amygdala sent much of its input to the amygdala's central nucleus that acted as a conduit for its main output pathways. Indeed, LeDoux was to show that all the components of the conditioned emotional response are mediated by different outputs of the central nucleus. Thus, lesioning the central pathway to the periaqueductal gray area of the midbrain interfered with the freezing response, whereas severing the pathway to the lateral hypothalamus abolished the increase in blood pressure. And a lesion made to the bed nucleus of the stria terminalis, which projects to the ventromedial hypothalamus, was later shown to stop the release of stress hormones involved in fear, including adrenocorticotropic hormone and corticosterone.

The fact that the medial geniculate nucleus has priority over the higher auditory cortex in fear-linked behaviour makes a great deal of sense. When faced with a dangerous situation, animals require an alarm mechanism allowing them to respond as quickly as possible to the event. The thalamic input to the amygdala provides this prioritised response. It takes around 12 msecs for a sound stimulus to reach the amygdala via the medial geniculate nucleus and at least twice as long via the auditory cortex. Thus, this arrangement enables fear processing to begin in the amygdala before we are aware of what is happening. Because failing to respond to danger is, for most animals, likely to prove fatal, it is safer to react first than waste crucial time analysing the event. Thus, reflexive emotional responses are evolution's solution to danger, and the subcortical route to the amygdala allows a "react first, think later" response to take place. In fact, it turns out that in humans, even the cortical sensory inputs reaching the amygdala are non-conscious – a finding confirmed in blindsight patients (i.e. those with damage to the visual cortex) and in experimental procedures that prevent conscious detection of events such as masking, where stimuli are presented too quickly to be "seen".

Despite this, cortical analysis still has an important role to play in assessing dangerous situations. For example, there may be instances when we need to voluntarily confront a fearful event rather than running away. Higher thought processes will also enable us to form a more accurate understanding of the danger and make decisions based on previous experience or reasoning. Such interpretation is most likely to involve the frontal lobes and presumably it bypasses the amygdala. Another important area of the brain involved in fear is the hippocampus. Although animals with hippocampus lesions still learn a conditioned emotional response, they do not show the normal signs of fear and anxiety when placed back into the environment where the aversive events had previously taken place. In other words, hippocampal lesioned animals do not recognise whether an environment is dangerous or not.

Because the amygdala is the site where information about the tones and shocks are processed, it is reasonable to hypothesise that it is also the place where the neural basis of fearful learning events takes place. The discovery that long term potentiation occurs in the amygdala provides support for this hypothesis. Moreover, LeDoux has shown that the acquisition of conditioned fear in rats is blocked by infusion of glutamate antagonists into the amygdala, which prevents the formation of LTP. Fear conditioning has also been shown to increase the levels of calcium inside amygdala neurons through glutamate activation of NMDA receptors – a cellular change that activates various protein kinases, in turn stimulating gene expression and protein synthesis, resulting in the long-term consolidation of memory.

In recent years, LeDoux has attempted to address some of the broader theoretical issues concerning emotions. One subject that has long interested psychologists is the question of what comes first – the behavioural response or the emotion. Common sense tells us, if we see a dangerous snake, we will experience fear and then run away. However, in 1884, the American philosopher William James suggested an alternative theory: instead of saying we run because of fear, he proposed we become fearful because we run. If this is the case, then the autonomic changes in the body come about, along with the resulting action, before the emotion is consciously experienced. Although this view has its critics, LeDoux's work with Pavlovian conditioning clearly lends it considerable support.

The propensity to run away from danger is shared by just about every living thing. Indeed, as LeDoux has pointed out, even a bacteria can move away from harm in its environment. But this still raises the problem of how the conscious feeling of fear comes about, and here LeDoux is more controversial. For many psychologists, emotions arise exclusively from an "old" part of the brain called the limbic system. But LeDoux disagrees, contesting that much of the limbic system is not involved in emotion (or Pavlovian conditioning). Hence for him, the concept of a limbic "emotional brain" is misleading. LeDoux also argues that while the amygdala may contribute to producing fear responses, it is too simple to regard it as the brain's fear centre. In fact, LeDoux believes that the brain mechanisms that give rise to conscious emotions are not fundamentally different from those that give rise to perceptual conscious experiences. Viewed in this way, emotions are higher order representations of lower order (presumably subcortical) events interpreted by cogni-tive processes in the cortex. Although LeDoux accepts he has not yet fully explained the "William James problem" of how conscious feelings come about, he has made some progress towards solving it.

LeDoux left Cornell in 1989, moving to New York University where he has remained since, rising through professorships, to take the position of university professor – the highest honour for a faculty member at NYU. He now heads the Emotional Brain Institute, founded in 2007, as well as directing the university's Centre for the Neuroscience of Fear and Anxiety. With over 30 years of research into emotion, and author of popular books including *The Emotional Brain*, LeDoux has long been at the forefront of bringing the subject of emotion in from the cold. Indeed, by trying to understand emotion in terms of information processing rather than a phenomenological experience, LeDoux has helped establish the field of affective neuroscience. He has also done more than any other investigator to make the amygdala a household word. In addition, LeDoux is the only neuroscientist in this book who is a lead singer and guitarist of a rock band (called "The Amygdaloids"). And his song writing is clearly influenced by his work. On the first track of his band's album, *Heavy Mental*, LeDoux sings "Why do we feel so afraid? Don't have to look very far. It's all in a nut—in your brain".

Notes

1 This is a highly simplified account of the auditory pathways of the brain and does not include the trapezoid bodies or the superior olivary complex but will suffice here.
2 Interestingly, medial geniculate nucleus (MGN)-lesioned rats do gradually become fearful in response to the conditioning – probably due to the cortex taking over the learning. Nonetheless, this learning is far slower.

LeDoux's major writings

LeDoux, J.E. (1991) Emotion and the limbic system concept. *Concepts in Neuroscience*, 2, 169–199.

LeDoux, J.E. (1994) Emotion, memory and the brain. *Scientific American*, 270, 50–57.

LeDoux, J.E. (1998) *The Emotional Brain: The Mysterious Underpinnings of Emotional Life*. New York: Simon and Schuster.

LeDoux, J.E. (2000) Emotion circuits in the brain. *Annual Review of Neuroscience*, 23, 155–184.

LeDoux, J.E. (2012) Rethinking the emotional brain. *Neuron*, 73, 653–676.

LeDoux, J.E. (2012) Evolution of human emotion: A view through fear. *Progress in Brain Research*, 195, 431–442.

LeDoux, J.E. (2016) *Anxious: Using the Brain to Understand and Treat Fear and Anxiety*. London: Penguin Books.

LeDoux, J.E. and Brown, R. (2017) A higher-order theory of consciousness. *Proceedings of the National Academy of Sciences*. www.pnas.org/content/114/10/E2016.short.

Further reading

Anderson, A.K. (2007) Feeling emotional: The amygdala links emotional perception and experience. *Social Cognitive and Affective Neuroscience*, 2, 71–72.

Kim, J.J. and Jung, M.W. (2006) Neural circuits and mechanisms involved in Pavlovian fear conditioning: a critical review. *Neuroscience and Biobehavioral Reviews*, 30, 188–202.

Maren, S. (2001) Neurobiology of Pavlovian fear conditioning. *Annual Review of Neuroscience*, 24, 897–931.

Panksepp, J. and Biven, L. (2012) *The Archaeology of Mind: Neuroevolutionary Origins of Human Emotions*. New York: Norton.

Whalen, P.J. and Phelps, E.A. (2009) *The Human Amygdala*. The Guildford Press: London.

Chapter 38

Patricia Churchland (b. 1943) and Paul Churchland (b. 1942)

Patricia and Paul Churchland are philosophers who espouse a strong from of reductionism called "eliminative materialism", which holds that all processes attributable to the mind will one day be explainable in terms of neurobiology – an advance that will require the cooperation of neuroscience and philosophy.

Patricia Smith Churchland was born in the small town of Oliver, British Columbia, the most westerly of Canada's provinces, and raised on a farm in the picturesque South Okanagan valley. Neither of her parents were university educated or strongly religious. Churchland's father spent most of his time working on the family farm, which had an orchard and some animals, while her mother was a nurse. Despite their limited education, both parents encouraged Patricia to do well at school and attend college, despite being the butt of jokes from their farmer neighbours who thought this was a waste of money. Churchland attended the University of British Columbia where she began studying psychology but increasingly turned to philosophy. It was here, in the second year of her study, she would meet in a class on Plato, her future husband Paul, although romance was not yet in the air. Graduating with a BA degree in philosophy in 1965, Churchland then studied at the University of Pittsburgh, obtaining her masters in 1966, before moving to England where she received her bachelor of philosophy (equivalent to a PhD) in 1969 from Somerville College, Oxford. Her dissertation was on whether human behaviour could be considered as protoscientific – a subject she later regarded as quite boring. Nor was it anything to do with the subject of neuroscience. The introduction to this subject would only come in the next stage of Churchland's career when she was hired as a junior professor in philosophy at the University of Manitoba in Winnipeg from 1969 to 1982. Here, she studied neuroanatomy and neurophysiology for the first time and became "totally hooked" when she saw brain cells under a microscope, visited patients in hospital with neurological problems and realised the sorts of issues that neuroscience could address. Somehow she managed to run classes in logic during this schedule – a period she described as a bit hectic but always exciting. Following this, Churchland spent a year as a visiting scholar at Princeton and then another year back at the University of Manitoba as full professor. She is currently a professor of philosophy at the University of California, San Diego, where she has taught since 1984, and adjunct professor at the nearby Salk Institute for Biological Studies since 1989. Two people at Salk who encouraged Churchland to develop the discipline of neurophilosophy are Jonas Salk, whose Institute bears his name, and Francis Crick, the co-discoverer of DNA and pioneer of consciousness research.

Paul Montgomery Churchland was born in Vancouver in 1942. His father was skilled in wood and metal work, taught science in a local high school, and founded a boat-works company, while his mother enjoyed sewing. As a young boy, Paul became obsessed with science fiction and decided on becoming an aerodynamical engineer since this might lead him to build flying saucers. With this ambition not totally extinguished, Churchland went to the University of British Columbia where he graduated with a BA in philosophy, physics and mathematics. From 1964 to 1967, he studied at the University of Pittsburgh, receiving his PhD under the supervision of philosopher Wilfred Sellars who was a strong advocate of critical realism. This is the theory that some of our sense data (e.g. solidity, motion, extension) are real "primary" qualities of the physical world, whereas other aspects of reality (e.g. vision, taste) are "secondary" or imagined qualities, which arise from mental representation. This theory greatly influenced Churchland, who developed a scepticism about intuitively plausible ideas that appeared straightforward but which were actually conceptual and invented – a view that later shaped his eliminative materialist approach to neuroscience. In 1966, Churchland became an instructor in philosophy at Pittsburgh, and it was around this time he visited Oxford where he begun a relationship with Patricia, whom he would marry in 1969 and have two children (Mark and Anne – both of whom would become neuroscientists). From then onwards, the couple would not only share similar philosophical viewpoints but also hold appointments at the same institutions. In 1984, Paul Churchland became a professor at the University of California, San Diego, where he served as department chair from 1986 to 1990. As of February 2017, he is also associated with consciousness studies in the philosophy department at Moscow State University.

Eliminative materialism is the claim that our ordinary understanding of mental phenomena is inaccurate and will one day be replaced (or eliminated) by a new vocabulary. Although it has a long history, there is little doubt that the Churchlands have done much to breathe new life into the theory and introduce it to a wider audience. This quest began in 1981 with a paper by Paul Churchland, which outlined several arguments for rejecting "folk" psychology. This is the common sense explanation of people's behaviour, which we tend to use in our dealings with others. For example, if we ask the question: why did John go to the local shop? We may get the reply: because he wanted to buy some milk. It is an explanation that attributes John's behaviour to his internal mental beliefs and desires. But Churchland claims that these everyday accounts of psychological states are misleading since they do not exist in any material sense. To illustrate his point, Churchland argues that "folk" attributions can be regarded as similar to phlogiston – an imaginary substance that up until the late eighteenth century was believed to be released during fire and combustion. And like the theory of phlogiston, which has been shown to be false by modern chemistry, folk psychology will one day be shown to be a misleading theory of mind by a mature neuroscience.

It is interesting to note that a similar argument was made by the philosopher Gilbert Ryle in his 1949 book *The Concept of Mind*, who, in attacking Cartesian dualism, described our common sense use of the word "mind" as a *category mistake*. To explain this, Ryle gave an example of a visitor who is shown around Oxford University. After viewing the colleges and library, he then asks "where is the university?" Clearly, the visitor is mistaken over the conceptual meaning of what constitutes a university, and in the same way, Ryle holds that our concept of mind in everyday language is also confused. In short, he argues that attributing the mind as a cause of behaviour is incorrect, since it has no physical existence. It is therefore a philosophical illusion or in Ryle's own words "the ghost in the machine".

The central point of the Churchlands argument is that when we speak of mental phenomena, we are actually referring to neurophysiological states consisting of neural activity (action potentials, neurotransmitter release, etc.) that is primarily involved in some type of functional and computational process. Thus it follows, the mind is simply what the brain does – and one day, the Churchlands believe, this will result in neuroscience replacing the imprecise generalisations of folk psychology with a new language of precise physical terminology. In doing this, neuroscience will show that there is just one kind of ontology[1] in our world – the physical one. The Churchlands also point out that similar eliminative reductions have taken place many times in the history of science (e.g. the recognition that temperature is nothing more than molecular kinetic energy) and there is every reason to believe tthat he same reduction will occur for mental states in neuroscience.

It should be noted that eliminative materialism is not the only form of physicalism that can be adopted by neuroscientists. An alternative approach, and one with wider support in the philosophy community, is *non-reductive materialism* – a form of monism that holds that while everything is made from material substances, the mental properties of the brain are causally efficacious and can not be explained in terms of lower level physical events. This view is also compatible with the idea that the mind is an emergent property of neural activity. Opponents of eliminative materialism also claim that full reductionism will never be able to account for the phenomenal features of our subjective experience and always leave out something which is crucial – namely, the feeling of what it is like to smell a rose, see the colour red or hear the screech of chalk on a blackboard. These qualitative characteristics of sensation are called *qualia*, and for many philosophers constitute one of the defining features of consciousness, which are impossible to clarify in physical terms. As might be expected, the Churchlands disagree and retort that one day we will be able to fully elucidate how qualia come about – even though they accept it may not necessarily result in the production of the phenomenon the theory explains (i.e. the smell of a rose).

In 1986, Patricia Churchland published her first book *Neurophilosophy*, which attempted to provide an interface between neuroscience and philosophy. Partly to persuade philosophers that to comprehend the mind one must understand the brain, and partly to introduce neuroscientists to the rich array of philosophical problems that they may one day contribute to solving, Churchland focused, in large part, on the thorny issue of the *mind-brain problem*. This is arguably the most perplexing and intractable issue facing both philosophy and neuroscience, and one going back to the ancient Greeks – although it was Descartes in the seventeenth century who provided its modern focus when arguing that mind and body are two separate things (a position known as substance dualism). He reached this conclusion by reasoning that the nature of the mind (a thinking, non-extended thing) is different from the body (an extended, non-thinking thing). Thus, for Descartes, it was possible for one to exist without the other. As Churchland points out, since Descartes, there have been two schools of thought: one embracing the idea of the mind as a spiritual non-physical thing and the other who believe that all mental activity is dependent on physical (biological) organisation.

Many philosophers are convinced that conscious processes will never be explained in neural terms and several famous "thought experiments" support their contention. Perhaps the most famous is one invented by American philosopher Thomas Nagel who asked a simple question: what is it like to be a bat? The point that Nagel tries to make is, no matter how much we learn about a bat, we will never know what it is like to be one. In other words, there is something that is, or feels like, to be a conscious thing – and neuroscience will never capture it. Another influential thought experiment was devised by Australian philosopher Frank Jackson, who invented a hypothetical character called "Mary", a colour-blind scientist, who knew every fact there was to know about the neural basis of colour vision. The question is: would she recognise a colour (say red) if she saw it for the first time? Jackson argues the answer is no, and he backs up this assertion by arguing that conscious sensation provides us with something that cannot be explained in scientific terms. Another strong advocate of dualism is David Chalmers, who asks: could a zombie exist who was exactly like you in every neural and behavioural respect, yet have no conscious experience whatsoever? For Chalmers, the answer is yes – and if this is the case, then it suggests to him that consciousness is an "extra ingredient" above and beyond the neural activity of the brain.

The Churchlands maintain that such thought experiments are based on misunderstandings and will one day be shown to be spurious. They are also confident that the identity between mental and brain states will one day be recognised as being exactly the same. Thus, in the event that a certain pattern of neuronal activity is shown to be identical to the conscious sensation of smelling a rose, then there will be no need for explaining why this is identical to the conscious sensation – it just is. But again, as always, there are philosophical objections. For example, imagine two people with identical brains – one having a visual system tuned to seeing red and the other to green. Is this inverted spectrum possible? Many neuroscientists would contend it is. But if so, would we ever be able to know which person is seeing red and the other green with our recording electrodes or fMRI machine? Some would say no.

Patricia and Paul Churchland have recognised that neurophilosophy has the potential to ask some deep questions about the biological basis of human nature, and because of this, there is every likelihood that the discipline will develop into a major branch of neuroscience in the near future. The Churchlands have also placed the mind-body problem at the forefront of neurophilosophical problems to be solved – cajoling neuroscientists to prove that all our mental processes, from thinking and feeling to identity, consciousness and free will, are produced by physical events in the brain – a situation that Francis Crick once called "the astonishing hypothesis". But is this a realistic aim? Maybe we can take some comfort from the fact that neuroscience is a young subject (as this book has shown). In fact, the Churchlands believe that our present knowledge concerning the brain is so paltry that even if presented with the answers, we would still not be able to comprehend the solution to many problems in neuroscience. Patricia Churchland makes this point by getting her students to imagine they are Albertus Magnus (a thirteenth-century German theologian and philosopher), who is wondering about the burning of wood and one night is visited by a Martian who whispers "Hey, Albertus, the burning of wood is really rapid oxidation!" But what could he do? He has no knowledge of structural chemistry, oxygen or the true nature of temperature. Put simply, the great intellect of Albertus Magnus would not be able to make any sense of the information he is given. It is a beguiling thought that as students of neuroscience, we may be in a similar position regarding our understanding of the human brain.

Note

1 Ontology is a term beloved by philosophers, which refers to the branch of metaphysics that deals with the nature of being (from the Greek "onto" meaning being and "logia" meaning study or spoken discourse).

Patricia and Paul Churchland's major writings

Churchland, P.M. (1981) Eliminative materialism and propositional attitudes. *Journal of Philosophy*, 78, 67–90.

Churchland, P.M. (1988) *Matter and Consciousness*. Cambridge, MA: MIT Press.

Churchland, P.M. (1996) *The Engine of Reason, the Seat of the Soul*. Cambridge, MA: MIT Press.

Churchland, P.M. (2012) *Plato's Camera*. Cambridge, MA: MIT Press.

Churchland, P.S. (1985) *Neurophilosophy: Towards a Unified Science of the Mind-Brain*. Cambridge Mass: MIT Press.

Churchland, P.S. (1996) The hornswogle problem. *Journal of Consciousness Studies*, 3, 402–408.

Churchland, P.S. (2002) *Brain-Wise: Studies in Philosophy*. Cambridge Mass: MIT Press.

Churchland, P.S. (2013) *Touching a Nerve*. London: Norton and Co.

Further reading

Bickle, J., Mandik, P. and Landreth, A. (2010) The philosophy of neuroscience. *Stanford Encyclopedia of Philosophy*. https://plato.stanford.edu/entries/neuroscience/.

Brook, A. and Akins, K. (2005) *Cognition and the Brain: The Philosophy and Neuroscience Movement*. New York: Cambridge University Press.

Burton, H. (2014) *Philosophy of Brain: A Conversation with Patricia Churchland*. Open Agenda Publishing (ebook).

Crick, F. (1994) *The Astonishing Hypothesis: The Scientific Search for the Soul*. New York: Touchstone.

Keeley, B. (2006) *Paul Churchland*. New York: Cambridge University Press.

McCauley, R.N. (1996) *The Churchlands and their Critics*. Oxford: Blackwell.

Ramsey, W. (2013) Eliminative materialism. *Stanford Encyclopedia of Philosophy*. https://plato.stanford.edu/entries/materialism-eliminative/.

Author index

Adrian, Edgar 37–43, 65, 66, 87
Akelaitis, Andrew 172, 248
Alzheimer, Alois 23, 24
Andersen, Per 220, 221
Aserinsky, Eugene 82–85
Aub, Joseph 115
Axel, Richard 226–232
Axelrod, Jules 111, 135–141

Bard, Philip 52
Barger, George 45
Barnes, Carol 244
Bartlett, Frederic 122
Behan, Peter 175
Berger, Hans 41, 84
Berlin, Rudolf 25
Bernard, Claude 157
Bernstein, Julius 200
Bielschowsky, Max 24
Bjorklund, Anders 167
Bliss, Timothy 76, 219–225
Bogen, Joseph 151, 152, 248
Boldrey, Edwin 32–33
Bonhoeffer, Karl 171
Booker, Barbara 146
Bremer, Frédéric 65, 68, 69
Bremer, Sydney 206
Brindley, Giles 206
Broca, Paul 61, 96–97
Brock, Lawrence 104
Brodal, Alf 219
Brodie, Bernard 136, 137, 164
Brodmann, Karl 23–29, 61
Bronk, Detlev 41
Brownstein, Michael 140
Buck, Linda 226–232
Bucy, Paul 123, 213
Bueker, Elmer 145
Burns, Benedict 221

Camis, Mario 65
Campbell, Alfred 27
Cannon, Walter 52
Carlsson, Arvid 163–169, 180, 196
Chalmers, David 265
Churchland, Patricia 261–267
Churchland, Paul 261–267
Clark, George 62
Cobb, Stanley 94
Cohen, Stanley 146, 147
Cole, Kenneth 86, 87, 90
Conway, Dulcie 243
Coombs, Jack 104
Creed, Stephen 101
Creutzfeldt, Otto 199
Crick, Francis 206, 262
Curtis, Howard 86, 87
Cushing, Harvey 31

Dale, Henry 21, 44–50, 102–103, 105, 109
Damasio, Antonio 175
Deecke, Luder 188
Déjérine, Jules 172
del Castillo, José 110
Dement, William 83
Descartes, Rene 14, 264, 265

Dickinson, William 17, 18
Dostrovsky, Jonathan 241
Du Bois Reymond, Emil 47
Dudley, Harold 48

Eccles, John 49, 75, 100–106, 108, 109, 110, 185, 207
Ehrlich, Paul 20
Elliott, Thomas 19–20
Erb, Wilhelm 9, 10
Ewins, Arthur 46

Falck, Bengt 167
Fatt, Paul 75, 108–109, 110
Feinstein, Bertram 185–186
Feldberg, Wilhelm 48–49, 178
Ferrier, David 53
Fick, Alfred 115
Finger, Stanley 42
Flourens, Pierre 61
Forbes, Alexander 38, 39
Foerster, Otfrid 31, 32
Forel, August 4
Foster, Michael 5, 16
Franz, Shepherd 59
Fritsch, Gustav 53, 61

Galaburda, Albert 175
Gall, Franz 61
Gardner-Medwin, Tony 222
Gaskell, Walter 17, 18, 44
Gasser, Herbert 86
Gazzaniga, Michael 152, 153, 172, 234, 247–253, 255
Gerald, Ralph 108, 184
Geschwind, Norman 170–176
Gibbs, Frederic 94
Gildermeister, Martin 107
Gilman, Alfred 178
Goldstein, Kurt 171, 172
Golgi, Camillo 2, 3, 4, 5, 6, 65, 143
Goltz, Friedrich 9
Goodale, Melvyn 216
Granit, Ragnar 101
Greengard, Paul 168, 177–196
Gregory, Richard 208
Gross, Charles 214
Grundfest, Harry 192

Hamburger, Victor 144, 145, 146
Hartline, Keffler 157, 178
Hebb, Donald 72–78, 123, 129, 206, 207, 213, 221
Hess, Walter 51–57
Hickok, Gregory 238
Hill, Archibald 107
Hillarp, Nils 165, 167
His, Wilhelm 4
Hitzig, Eduard 53, 61
Hodgkin, Alan 75, 86–92, 108, 109, 178, 199, 200, 202
Hokfelt, Tomas 167
Horsley, Victor 32
Hortega, Pio 31
Hubel, David 154, 156–162
Hughlings Jackson, John 34, 61
Huxley, Andrew 75, 86–92, 108, 109, 178, 199, 200, 202

Ito, Masao 208
Iwai, Eichii 215

Jackson, Frank 265
James, William 258
Jasper, Herbert 32, 156

Kandel, Eric 168, 191–197, 228
Kaplin, Edith 173
Katz, Bernard 75, 89, 102–103, 107–113, 139, 199, 200
Kebabian, John 180
Kety, Seymour 114–138
Kleitman, Nathanial 79–85
Kluver, Heinrich 123, 213
Koch, Robert 9
Kornhuber, Hans 188
Koskinas, Georg 27
Kraepelin, Emil 26
Krebs, Edwin 178
Kühne, Wilhelm 5
Kuffler, Stephen 103, 157, 158, 159, 192
Kupfermann, Irving 194
Kuypers, Hans 214

Langley, John 9, 16–22, 44, 47, 52

Lashley, Karl 58–64, 73, 77, 125, 150, 151, 172, 213, 214, 215, 248
LeDoux, Joseph 98, 254–260
Lerner, Aaron 139
Levi, Giuseppe 143
Levi-Montalcini, Rita 142–148
Levitsky, Walter 175
Libet, Benjamin 184–190
Liepmann, Hugo 173
Lindsley, Donald 66, 69
Lindvall, Olle 167
Lindvist, Margit 166, 180
Ling, Gilbert 108
Livingstone, Margeret 161
Loewi, Otto 47, 48, 49, 102
Lorente de No, Raphael 74
Lømo, Terje 76, 219–225
Lucas, Keith 37, 38
Lugaro, Ernesto 6
Lux, Hans 199

MacLean, Paul 93–99
Magoun, Horace 65–71
Marr, David 205–211
McNaughton, Bruce 243
Melzack, Ronald 241
Mesulam, Marsel 175
Meyer, Hertha 145
Miledi, Ricardo 111–112, 200
Miller, George 252
Milner, Brenda 121–127, 192
Milner, David 216
Milner, Peter 122–124, 126, 129–132
Mishkin, Mortimer 123, 212–218
Moniz, Egas 55
Morris, Richard 243, 244
Moruzzi, Giuseppe 65–71, 220
Moser, Edvard 244–245
Moser, May-Britt 244–245
Myers, Ronald 151

Nadel, Lynn 242
Nagel, Thomas 265
Neher, Erwin 198–204
Nissl, Franz 23, 24

O'Keefe, John 240–246
Olds, Marianne 132
Olds, James 128–134

Pankseep, Jaak 133
Papez, James 95–96
Pavlov, Ivan 59–60, 61, 72
Penfield, Wilder 30–36, 61, 73, 123–124
Piéron, Henri 80
Popper, Karl 105
Pribram, Karl 61, 123, 213, 215
Purkinje, Jan 207

Quadfasel, Fred 171, 172

Ramachandran, Vilayanur 238
Ramón y Cajal, Santiago 1–7, 9, 25, 207
Ranck, James 244
Ranson, Stephen 66
Rasmussen, Theodore 33, 126
Ressler, Kerry 229
Richardson, Bruce 81
Rizzolatti, Giacomo 233–239, 250
Rose, Caleb 8
Rosvold, Haldor 213, 214
Ryle, Gilbert 263

Sakmann, Bert 198–204
Salk, Jonas 262
Sato, Takaaki 229
Schleiden, Matthias 3
Schmidt, Carl 115
Schmidt, Francis 171
Schwann, Theodor 3
Schwartz, James 194–195
Scoville, William 124–125, 192
Sellars, Wilfred 262
Sherrington, Charles 5, 8–14, 19, 30, 40, 41, 72, 100–101, 103
Silverstein, Saul 227
Singer, Marcus 171
Skinner, Burrhus 73, 192
Smith, William 248
Snyder, Solomon 140
Sokoloff, Louis 117, 119
Solomon, Richard 129

Spencer, Alden 192
Sperry, Roger 149–155, 173, 234, 248–250
Sutherland, Earl 178, 179
Symonds, Charles 171

Tauc, Ladislav 193
Thompson, Robert 255

Ungerleider, Leslie 212–218

Van Wagenen, William 248
Verworn, Max 52
Virchow, Rudolf 9
Vogel, Philip 152
Vogt, Oskar 23–24
Von Economo, Constantin 27
Von Euler, Ulf 111, 139
Von Gerlach, Joseph 3
Von Helmholtz, Hermann 12, 208
Von Kölliker, Albert 4

Waldeyer, Wilhelm 4
Walter, Grey 205
Watson, John 58, 59, 60
Weigert, Carl 23
Weiskrantz, Larry 250
Weiss, Paul 150
Weissbach, Herbert 139
Wellcome, Henry 44, 45
Wenger, Daniel 189
Wernicke, Carl 174, 175
Westphal, Carl 9
Wiesel, Torsten 154, 156–162
Wigler, Michael 227
Wong-Riley, Margeret 160

Young, John 87–88, 104

Zangwill, Oliver 122
Zotterman, Yngve 40, 41
Zou, Zhihua 231

Subject index

acetylcholine 19, 21, 46–49, 102, 109–110, 111, 138, 201, 202, 224
acetylcholinesterase 48, 138
action potential 40, 87–91, 111
adrenaline (epinephrine) 19, 138, 178
adrenal glands 19, 45, 165, 167
alexia 172
alkaloids 17
all-or-nothing law 37–40
Alzheimer's disease 24, 147, 245
amnesia 124–126
amygdala 124, 125, 132, 172–174, 228, 254–259
aphasia 171–174, 228
Aplysia 193–196, 228
apraxia 173
ascending reticular formation 69–70, 132
association areas of cortex 174–175, 214
atropine 17, 46, 28
auditory cortex 256–257
autonomic nervous system 17–21, 96
autoreceptors 168
axons 4, 88–90

behaviourism 60–63, 73, 150
blindsight 250
Broca's area 27, 33, 235
Brodmann's areas 25–28

calmodulin 180, 233
catechol-0-methyltransferase (COMT) 138
cell assembly 74–75
cerebellum 2–3, 42, 65, 97, 206–208
cerebral blood flow 115–117, 119
cerebral cortex (neocortex) 25–28, 32–34, 60–63, 67–68, 96, 97, 151–154, 172–175, 208, 214–217, 235–237
chlorpromazine 166, 180
cingulate gyrus 95, 132, 237
circadian rhythms 81, 82, 139, 164
classical (Pavlovian) conditioning 59–60, 255–258
cognitive mapping 241–245
cognitive neuroscience 252
consciousness 125, 188–190
contralateral neglect 251
corpus callosum 150–154, 172–175, 215, 234, 248
cresyl violet staining 24, 28
curare 17, 46, 109
cyclic adenosine monophosphate (cAMP) 178–179, 195, 196, 228
cyclic adenosine monophosphate regulated phosphoprotein (DARPP-32) 181
cyclic adenosine monophosphate response element binding protein (CREB) 196

diencephalon 52–55
disconnection syndromes 172–175
dopamine 133, 139, 165–168, 179–181, 196

electrical recording of nerve cells 104, 108, 133, 156–160, 220–224, 234–237, 241–245
electrical recording of nerve fibres 38–41, 88–91
electrical stimulation of the brain 32–35, 52–55, 67–68, 130–133, 186–187, 220–222
electroencephalography (EEG) 41, 68–69, 84, 185
eliminative materialism 263–266
emotion 52–55, 94–98, 153, 255–259
enzymes 137
epilepsy 31–35, 94
ergot 45
eserine 48, 103, 109
excitatory postsynaptic potential 104

fear 252, 255–259
free will 187–188
frequency coding 38–42
frontal cortex 33, 73, 96, 126, 132, 174, 213, 230, 234, 257
functional brain scanning techniques 117, 126, 236, 237

gamma aminobutyric amino acid (GABA) 139, 182
giant squid axon 86–88, 111, 171
glutamate 139, 181, 182, 194, 195, 223, 223, 258
Golgi stain 2, 143
grid cells 244

haloperidol 168, 180
Hebb's rule 75, 207
hippocampus 94, 95, 124–126, 132, 192–193, 208, 220–223, 240–245, 258
histamine 45–46, 47, 48
hypothalamus 52–55, 66, 68, 70, 82, 95, 132–133, 257

inferotemporal cortex 123, 214–215
inhibition 10, 12, 13, 101–104, 158, 192, 207
inhibitory postsynaptic potential 104
ions and ion channels 88, 89–91, 109–111, 181, 182, 195, 199–203

l-Dopa 165, 166
lesioning techniques 11, 53, 60–63, 68, 124–125, 151, 213–217, 248, 256–257
limbic system 96–98, 132, 230, 258
localisation versus holistic debate 28, 61–63, 73, 133, 172–174
long term potentiation 76, 220–224, 258

mass action 61–63
medial forebrain bundle 132
medial geniculate nucleus 256
melatonin 82, 112, 139
memory 34–35, 60–63, 74–76, 124–126, 193–196, 213–216, 222–224, 242, 258
mind-body problem 14, 35, 56, 63, 100, 105, 154, 263–266
mirror neurons 234–238
motor cortex 24, 27, 32–33, 67, 188, 234–236
muscarinic receptors 21, 47, 48

nerve growth factor 144–147
neuromuscular junction 5, 20, 46, 48, 49, 103–105, 108–110, 138
neuron doctrine 4
neurophilosophy 261–267
nicotine 12, 14, 32
nicotinic receptors 21, 47, 48
nigral-striatal pathway 167
nitrous oxide 224

NMDA receptor 223–224, 258
noradrenaline (norepinephrine) 19–21, 45, 138–139, 165, 167
nucleus accumbens 132, 133, 181

odorant receptors 228–231
olfaction 42, 95, 96, 228–231

Papez circuit 95–96
parasympathetic nervous system 18–21, 46, 47, 49, 52, 55
parietal lobes 33, 42, 186, 216, 236, 251
Parkinson's disease 138, 166, 168
patch clamp method 201–202
pineal gland 112
place cells 241–243
planum temporale 175
positron emission tomography (PET) 118, 119, 126, 236
premotor cortex 33, 234, 235
proprioception 11, 24
protein kinases 223, 258
Prozac (fluoxetine) 167, 181
Purkinje cells 3, 207, 208

quantal release of neurotransmitter 110–111

receptor(s) 19–21, 45–49
reserpine 164, 165
reticular formation 67–69, 130, 131

schizophrenia 117, 118, 166–167, 168, 213
second messengers 178–181
self-stimulation of the brain 130–133
serotonin 139, 164, 165, 167, 195, 196
sleep 54, 55, 79–84
somatosensory cortex 24, 27, 33, 42, 186–187
sparks versus soup controversy 102–103
split-brain studies 150–154, 248–252
striatum 166, 167, 180, 181
superior cerebellar peduncle 67, 68
superior cervical ganglion 102, 179–180
sympathetic nervous system 18–20, 45, 47, 49, 52, 102, 139, 146
synapses 5, 12, 18, 20–21, 47, 75–76, 101–105, 108–111, 192–194, 196, 207, 208, 219, 221

temporal lobes 34, 73, 123–125, 174, 192, 213–216, 230, 236
thalamus 52–55, 67, 68, 69, 70, 94, 95, 130, 187, 230, 256
tricyclic antidepressants 139, 167

unconscious processing 188–189, 216, 250–252, 255, 257

vesicles 110, 167, 181
vision 208–210, 214–217
visual cortex 158–161

Wada test 126
Wernicke's area 33

For Product Safety Concerns and Information please contact our EU representative GPSR@taylorandfrancis.com
Taylor & Francis Verlag GmbH, Kaufingerstraße 24, 80331 München, Germany

www.ingramcontent.com/pod-product-compliance
Ingram Content Group UK Ltd.
Pitfield, Milton Keynes, MK11 3LW, UK
UKHW020130200926
475702UK00005B/687
9781138576124